Optical Fiber and Applications

Edited by Thamer A. Tabbakh

Published in London, United Kingdom

Optical Fiber and Applications
http://dx.doi.org/10.5772/intechopen.104032
Edited by Thamer A. Tabbakh

Contributors
Mengyuan Huang, Ansheng Liu, Kelly Magruder, Yann Malinge, Parastou Fakhimi, Hao-Hsiang Liao, David Kohen, Gregory Lovell, Wei Qian, Kiyoung Lee, Carsten Brandt, Mahtab Hakami, Yen-jung Chen, Erin Carabajal, Erle Guillermo, Seth Slavin, Deborah Glosser, Layla Russell, Paul Striby, Aishwarya Dash, Aruna Tripathy, Mohamed Jabha, Abdellah El Alaoui, El Houssine Mabrouk, Abdellah Jarid, Asiya E. A., Michael U. Onuu, Rufus C. Okoro, O. Enendu Uche

First published in London, United Kingdom, 2023 by IntechOpen
IntechOpen is the global imprint of INTECHOPEN LIMITED, registered in England and Wales, registration number: 11086078, 5 Princes Gate Court, London, SW7 2QJ, United Kingdom

British Library Cataloguing-in-Publication Data
A catalogue record for this book is available from the British Library

Additional hard and PDF copies can be obtained from orders@intechopen.com

Optical Fiber and Applications
Edited by Thamer A. Tabbakh
p. cm.
Print ISBN 978-1-83768-670-4
Online ISBN 978-1-83768-671-1
eBook (PDF) ISBN 978-1-83768-672-8

Meet the editor

Dr. Thamer Tabbakh is a distinguished associate professor specializing in electrical engineering, optics, photonics, and semiconductors. He is currently the general manager of the Microelectronics and Semiconductors Administration, King Abdulaziz City for Science and Technology (KACST), Saudi Arabia. With a robust academic background, Dr. Tabbakh attained his advanced degrees from the esteemed University of Central Florida College of Optics and Photonics (CREOL-UCF), USA, renowned as one of the foremost institutions for optics and photonics education on a global scale. Throughout his illustrious career, Dr. Tabbakh has made remarkable contributions to the field, evident through his extensive research portfolio. With more than 13 years of experience in the semiconductor industry, he has excelled both in research and administrative roles. Prior to his current position, Dr. Tabbakh served as the manager of the Nanotechnology and Semiconductors Center at KACST, further solidifying his expertise and leadership in the field.

Contents

Preface

In an age characterized by ever-advancing technology, the world finds itself intricately interconnected, and at the heart of this remarkable transformation lies the extraordinary medium of optical fiber. The slender strand of glass or plastic, with its ability to transmit vast amounts of information at the speed of light over long distances, has become the backbone of our modern communication infrastructure.

In this book, *Optical Fiber and Applications*, we embark on a captivating journey through the fascinating world of optical fiber and its diverse range of applications. Organized into five chapters, each delving into a specific aspect of the technology, this book provides a comprehensive understanding of the principles and real-world implementations of optical fiber.

Chapter 1, "Glass Fiber Waste from Wind Turbines: Its Chemistry, Properties, and End-of-life Uses" explores a crucial concern in sustainable energy generation. It delves into the chemistry and properties of glass fiber waste generated by wind turbines and analyzes the various end-of-life options for managing this waste effectively.

Chapter 2, "Spectral Characterization and Analysis of Underground Optical Fibre Cable Network Using Optical Time Domain Reflectometry", discusses underground optical fiber cable networks. Using optical time domain reflectometry, the authors perform spectral characterization and analysis to gain insights into the performance and condition of these networks.

Chapter 3, "A Comparative Analysis of Linear Modulation Schemes in the Presence of FLI in Wireless Optical Systems", presents a detailed comparative analysis of linear modulation schemes utilized in optical fiber systems. Through in-depth research and analysis, the chapter explores the strengths and limitations of different modulation techniques, shedding light on their potential applications and performance characteristics.

Chapter 4, "The Effect of Polymerization of 2.7-Divinylcarbazole-BenzoBis-Thiadiazole on Optical Fiber Properties", examines the impact of polymerization on optical fiber properties. It provides information on how the polymerization process of 2.7-divinylcarbazole–benzo-bis-thiadiazole affects the performance and characteristics of optical fibers.

Finally, Chapter 5, "Germanium on Silicon Avalanche Photodiode for High-Speed fiber Communication", investigates a cutting-edge technology for high-speed fiber communication. It reviews the potential of germanium on silicon avalanche photodiodes and explores their capabilities and applications in pushing the boundaries of fiber communication.

Throughout this book, we endeavor to provide a comprehensive and insightful exploration of optical fiber and its applications. We draw upon the collective knowledge and expertise of our authors to present a balanced blend of theoretical foundations, practical insights, and emerging trends.

We extend our gratitude to the researchers, engineers, and pioneers who have contributed to the development and advancement of optical fiber technology. Their dedication and innovation have shaped the landscape of modern communication and opened up new possibilities for connectivity.

We also express our appreciation to the readers who embark on this enlightening journey with us. Your curiosity and enthusiasm for understanding optical fiber and its applications serve as the driving force behind the creation of this book.

As you delve into the pages that follow, we invite you to immerse yourself in the captivating world of optical fiber, exploring its fundamental principles, diverse applications, and its potential to shape the future of communication.

Join us on this enlightening journey through the captivating world of optical fiber.
To R&R.

Thamer A. Tabbakh
GM of Microelectronics and Semiconductors,
King Abdulaziz City for Science and Technology,
Riyadh, Saudi Arabia

Chapter 1

Glass Fiber Waste from Wind Turbines: Its Chemistry, Properties, and End-of-life Uses

Deborah Glosser, Layla Russell and Paul Striby

Abstract

Glass fiber and glass fiber-reinforced polymers are of interest to engineers for a wide variety of applications, owing to their low weight, high relative strength, and relative low cost. However, management of glass fiber waste products is not straightforward, particularly when it is part of a composite material that cannot be easily recycled. This is especially the case for physically large structures such as wind turbine blades. This chapter deals with the challenges of managing this growing waste stream and reviews the structure and chemistry of glass fiber and glass fiber-reinforced polymers used in wind turbine blades, the separations processes for extracting the glass fiber from the thermoset resin, and end-of-life options for the materials. Thermodynamic evidence is reported and evaluated for a novel end-of-life solution for wind turbine waste: using it as a supplementary cementitious material.

Keywords: glass fiber, GFRP, cement, thermodynamics, wind turbines waste

1. Introduction

Glass fiber reinforced polymer (GFRP) is manufactured from the high temperature conversion of raw materials into a homogenous melt which is fabricated into fiber and bonded to a polymeric resin [1]. Raw materials used for the glass fiber (GF) portion are generally derived from geologic sources and may include silica sand [2]; clay [3]; calcite [4]; and borate minerals [5]. GFRP is used in multiple applications including textiles, electronics, aerospace, boats, and wind turbines, to name a few [6]. Despite the advantages of GFRP composites – such as low weight, high compressive strength, and relative low cost – the thermoset nature of the polymeric matrix presents a challenge for recycling of these materials, as the resin is not easily depolymerized from the original constituents and tends to undergo a loss of mechanical strength following separation from the resin [7]. This chapter describes the physical and chemical properties and use of GFRP with a specific focus on wind turbine blades, and evaluates current and prospective end-of-life strategies for dealing with this complex waste stream. Also discussed in this chapter are currently available recycling and end-of-life solutions for GF and GFRP, as well as a review of the separations processes for depolymerization of the resin. This chapter also recapitulates and discusses the results of a study on its prospective revalorization as a supplementary cementitious material (SCM). The strategic re-use of wind turbine GF in cement has the potential to not only to deal with this challenging waste stream, but also to offset the carbon emissions from cement clinkering [8].

1.1 Chapter objectives

This chapter is motivated by the need to understand limitations of current end of life options for GFRP used in wind turbine blades, and to systematically characterize the materials (chemical composition, structure) and technologies used in producing and processing wind turbine GFRP so that an assessment of its end-of-life uses as an SCM and otherwise may be presented End of life options for GFRP based materials are complicated by the thermoset nature of the composite material [9]. For this reason, repurposing of as-received GFRP-based parts is often the most economical solution. This makes the waste management from wind turbines – particularly their blades challenging - as the parts are physically quite large: they have an average rotor diameter of 127.5 meters and weigh on average 22,000 kg for the blades and hub alone [10]. Furthermore, the average lifespan of a wind turbine is 20 years, and by the year 2050, there is predicted to be over 40 million tons of GFRP based wind turbine materials entering the waste stream [9]. Although wind power is a lower-carbon option for energy generation than hydrocarbons, strategies are needed to deal with the

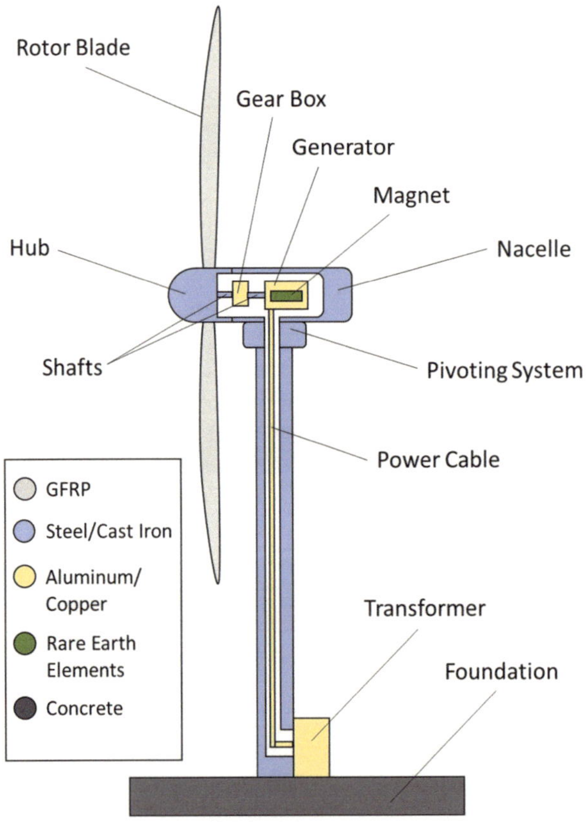

Figure 1.
Schematic diagram of wind turbine showing its parts and composition.

growing waste stream. One such strategy is recycling the GF portion of wind turbine waste for use as a supplementary cementitious material (SCM) [8]. While GFRP has previously been used by the cement industry for several applications it has not been commercially used as an SCM. This chapter will discuss the viability of using GF as an SCM based on a previously reported study, and will also describe the chemistry and structure of GFRP in relation to the current end-of-life options for the materials.

1.2 Review of wind turbine and GFRP anatomy

A wind turbine is a device that extracts energy from wind and converts it to mechanical energy, which is further converted into electricity by the wind turbine generator. Structurally, wind turbines are comprised of rotor blades attached to a base. The nacelle is the housing attached to the base, which stokes the turbine's generator. The anatomy of a wind turbine is shown in **Figure 1**, along with a key showing the primary materials used in each component.

As seen in **Figure 1**, the primary material of the rotor blades is GFRP, whereas concrete, rare earth elements, and e-metals and steel are the main materials of the remainder of the turbine. Of the GFRP in the blades, roughly 60% of the mass is comprised of GF, and the remaining 40% is comprised of resin. This chapter is primarily concerned with the turbine blade GFRP.

Figure 2 shows a colorized backscattered scanning electron microscope (SEM) image of GFRP, magnified x267. The yellow filaments apparent in the image are the glass fiber components, and the smooth green plate-like structures are the resin. Visual inspection of the orientation of the filaments in the resin material shows that the glass fibers are reinforcing the resin and are primarily oriented along the sagittal

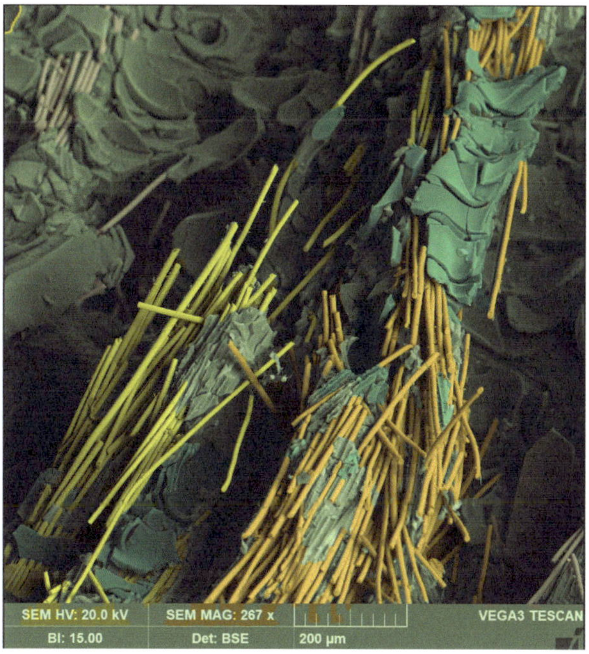

Figure 2.
Colorized backscattered scanning electron microscope image of GFRP.

plane. This dispersion and orientation contribute to the high tensile strength of GFRP materials. The high surface area to volume ratio of the GF is also apparent upon visual inspection. Compositionally, the glass fibers are derived from geologic materials that have been melted and processed into glass filaments, and the resin is a thermoset epoxy-based material. The SEM image reveals the nature of the composite material: that is to say, the degree to which the GF filaments are integrated into the resin. This is a helpful frame of reference when considering the technologies and methods available to separate the composite.

2. Review of current wind turbine GF and GFRP end-of-life options

End-of-life options for wind turbines generally fall under two categories: repurposing of old blades and wind turbine parts, and separation and recycling of the materials. Recycling processes require either the mechanical crushing of whole GFRP, or the extraction of specific materials from the GFRP and resin components in the composite, whereas repurposing uses the composite material with minor modifications in the as-received state. A waste hierarchy is conceptually shown in **Figure 3**, which describes the order of preference for the currently available methods to reduce the environmental impact of GFRP waste from wind turbine blades. While the design of materials to extend the service life of the parts is shown to be the least environmentally impactful option, repurposing and recycling represent intermediary options between life-extending design and landfilling/disposal. Repurposing and recycling of wind turbines are described below, along with a critique of the limitations and benefits of each.

2.1 End of life: repurposing

Repurposing of wind turbine parts by making minor modifications to their parts in their as-received state is the most common end of life option, due in large part to the higher cost of recycling [11]. Examples of repurposing include transforming the

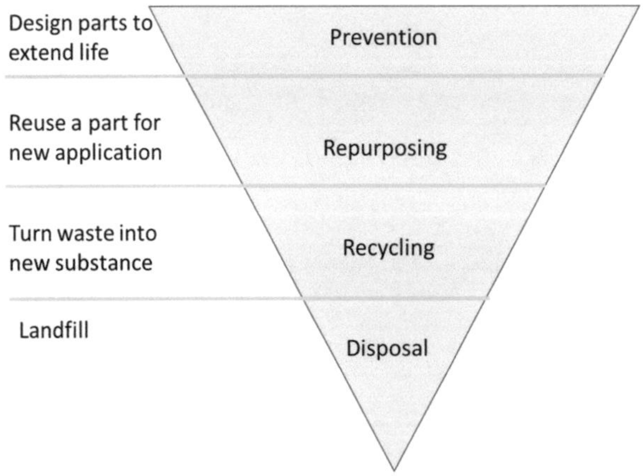

Figure 3.
Waste hierarchy showing order of preference for available methods to reduce the environmental impact of GFRP waste. Adapted from Nagle et al.

blades into load-bearing structural systems; cutting the blade for use in roof frames for small houses and/or affordable housing communities [12]; using the cut root section as a foundation for homes along flood plains [13]; and using the shear webs that connect the top and bottom halves of the blade as doors, window covers and more [14]. Other reuse options include use in community structures such as playgrounds, benches, bridges, and transmission poles. The defining feature of repurposing is that turbine parts are reused in their whole state, with the only modifications being cutting of the sections into smaller parts.

2.2 End of life: recycling

Although repurposing of wind turbine parts is the most economical and therefore common end-of-life option [11], recycling of the blades offers additional opportunities. Recycling refers to the mechanical crushing and reuse of the GFRP or the extraction of GF fibers from the composite for use as feedstock for other processes. Mechanical recycling, whereby scrap composites are grinded into smaller pieces and sieved based on particle size, is currently one of the few processes with current commercial operations, given the low cost and energy expense associated with it [15]. In a mechanical recycling process, the composite material is reduced by successive cutting, grinding, and then shredding. The resulting materials are reclaimed as either aggregate, powder, or clusters of GF. Some uses for mechanically recycled GFRP include use as a 3D printing feedstock; production of laminates; and reuse in new composites [15]. Current uses of recycled materials from these methods are limited by the presence of contaminants such as metals, mineral oils, and paints in the scrap, therefore it is imperative that the use cases for scrap recycling be agnostic to such impurities.

Although GF and GFRP have not been used as an SCM to date, the use of recycled GFRP in the cement industry is not without precedent: Mechanically recycled GFRP has been used as an aggregate replacement in concrete [16]. When used as an aggregate in concrete, tensile, flexural strength and shrinkage of the concrete has been found to improve, However, a review of several studies reveals that the influence is inconsistent, and varies depending on the chemical composition of the feedstock GFRP; the physical properties of the GFRP components; the presence of impurities in the materials; and imperfect dispersion of the materials in the concrete [17]. It should be noted that concrete aggregates are fundamentally distinct from SCMs: In the former case, granular materials ranging in sizes from roughly 10-40 mm are added to cement paste to form concrete. While the aggregate may chemically react with the cement paste, its main purpose is to improve the mechanical strength of the mixture. SCMs, on the other hand, are very fine ground particles that are added to the cement clinker (powder) after it is removed from the kiln, and participate in a series of chemical reactions to form reaction products which favorably impact the hardened cement paste. This is further described in Section 3 and demonstrated conceptually in **Figure 4**, which shows the differences between cement powder, SCM, aggregates, and concrete.

Recycling of GFRP may also refer to the extraction of GF fibers from the composite: due to the high energy cost and infrastructure requirement for this process, it has not been widely adopted [11]. However, another precedent for use of recycled wind turbine GFRP is cement co-processing. A small number of plants in Europe have been built specifically for the co-processing of wind turbine waste as a cement clinker replacement [18]. In a co-processing operation, the GFRP from wind turbines is

Figure 4.
Conceptual diagram showing difference between cement powder, SCM, paste, aggregate, and concrete.

shredded and then mixed with a solid recoverable fuel (SRF), about 50% of which is SFR and 50% GFRP by mass [19]. The resin portion of the GFRP in the wind turbine blade is used as fuel to heat the cement kiln, which offsets a portion of fossil fuel that would otherwise be used for this purpose. The clinker (powder) material in the cement feedstock is then replaced by the GF portion of the which reduces raw material consumption [8, 20].

2.3 Review of separations processes of glass fiber from polymer/composite material

As discussed in Section 2.2, recycling of the GF component of GFRP requires that the materials be separated from the matrix. Since the matrix in these composites often consists of thermoset resin which is not easily depolymerized (**Figure 2**), separation of the fibers from the resin can be an energetically expensive process [20]. The reader is referred to **Figure 2** for a visual examination of the physical integration of the glass fibers in the resin of a typical GFRP. Separations processes to extract the GF from the resin fall into two categories: chemical and thermal. The following sections describe thermal and chemical separations processes to elutriate GF materials from the resin. The limitations described below (energetic expense; presence of impurities; low yields) are indicative of why recycling of GFRP falls below repurposing in the waste management hierarchy (**Figure 3**), although it should be noted that technological advances may tip these balances in coming years.

2.3.1 Thermal separation

Currently, fluidized bed separation and pyrolysis are the only available thermal processes to accomplish separation of the GF from the composite, and the high operation temperatures required can damage the GF and negatively impact its mechanical properties [15]. This is critical to consider when evaluating how the extracted materials may be used in subsequent applications. The intended use of the resin must also be considered as it is a thermoset material cannot be usefully reshaped upon separation. Hence, any reuse of the resin must not require such a process.

In a fluidized bed process to separate the components, ground scrap composites are fed into a fluidized bed and heated to approximately 450 C with a velocity air jet, which causes the resin to volatilize and decompose. The GF can then be elutriated and sieved [21]. Organic contaminants that did not volatilize with the polymer must be removed in a bed grading process. Oxygen is required for this process in order to minimize formation of char. Fiber yields from fluidized bed processes are roughly 40% of the total fiber product. This is to say that approximately 60% of the feedstock fibers are not successfully rescued from current separations processes, which is a considerable limitation when considering the environmental and economic scalability. Technological improvements to increase the recovery of GF materials from fluidized bed separation are needed to make this process more efficient.

Pyrolysis is the second thermal separation process for extracting fibers from resin. In a pyrolysis process to thermally separate the components, GFRP composites are heated in an anoxic environment, which causes the matrix to decompose. This process involves the destructive distillation of the matrix, and produces hydrocarbons which can be harvested for their chemical value. Currently pyrolysis has only been proven at the laboratory scale and does not have any known commercial operation [21]. This is predicted to change in coming years as research and demand grow.

2.3.2 Chemical separation

In a chemical separation process for recycling GF components, the GFRP composite's polymetric matrix is dissolved into a chemical solution. Methods such as hydrolysis and glycosis are used to recover the raw materials. In a chemical process, glass fibers are extracted after the polymer is broken down into basic monomers. Nitric acid [7], sulfuric acid [22], and alternate acids can be used as a solvent. Glycol, methanol, ethanol, 1-propanol, and acetone may also be used in solvolysis to depolymerize the matrix and extract the GF [11]. The chemical yields from these processes are quite low, and the solid residue is often contaminated with organic material which requires further treatment. These are similar to the limitations of the pyrolysis process described in the previous section, and therefore, further research and investment is needed for these operations to become commercially valuable.

3. Review of cement and SCM chemistry

The previous sections detailed current end of life options for wind turbines, as well as the processes that can be used to separate the GF from the resin so that it can be recycled. The strategic revalorization of GF as a SCM requires that 1) the GF components be reactive with cement clinker, and 2) that they contain the chemical constituents need to form the primary reaction products (hydrates) of cement [23]. To that end section provides an overview of the chemistry of cement clinker, and the major reaction mechanisms and hydrates in cement. The chemistry of wind turbine GF specifically is then described and compared to the chemical constituents used in existing SCM in Section 3.2.

Ordinary Portland cement (OPC) clinker is derived from geologic materials (primarily clay and limestone). The clinker is produced by heating raw materials in a kiln at temperatures of more than 1450°C, where aluminum and iron-based oxides are

Chemical name	Formula	Notation	Name
Tricalcium silicate	Ca_3SiO_5	C_3S	Alite
Dicalcium silicate	Ca_2SiO_4	C_2S	Belite
Tricalcium aluminate	Ca_3AlFeO_6	C_3A	Aluminate
Tetracalcium aluminoferrite	Ca_2AlFeO_5	C_4AF	Ferrite
Calcium hydroxide	$Ca(OH)_2$	CH	Portlandite or CH
*Calcium silicate hydrate	$C_3S_2H_3$	C-S-H	C-S-H

Table 1.
*Cement chemistry notation (*note that C-S-H is non- stoichiometric).*

added as fluxing agents [24]. The primary constituents of OPC clinker are calcium silicates, ferrite, and calcium aluminates, with lesser amounts of potassium and sodium sulfates and magnesium oxide [24–26]. Cement chemistry notation is given in **Table 1**.

The reaction of these constituents in the cement pore solution yield solid hydration products such as calcium-silicate-hydrate (C-S-H) and calcium hydroxide (CH), the former of which gives cement its strength, and the latter of which helps to maintain the alkalinity of the pore solution [27, 28]. The alkalinity of cement pore solution is crucial to controlling deleterious processes such as corrosion and alkali-silica reaction [29]. To that end, the concentration of these elements in pore solution – and its pH – are of great importance when designing cement mixtures [30].

3.1 Supplementary cementitious materials

SCMs, when used in conjunction with OPC, contribute to the beneficial properties of the hardened cement by reacting hydraulically, and/or through the pozzolanic reaction (Eq. 1) in geochemical notation and Eq. (2) in cement chemistry notation to form desirable reaction products and pore solution chemistry.

$$Ca(OH)_2 + H_4SiO_4 \rightarrow CaH_2SiO_4 \cdot 2\,H_2O \tag{1}$$

$$CH + S \rightarrow C - S - H \tag{2}$$

SCMs are generally added to the clinker after the kilning process, and before the addition of water to the mixture (**Figure 4**). Materials generally used as SCMs include fly ash, silica fume, ground blast furnace slag, and metakaolin, although any material with appropriate chemical composition and reactivity may be used for this purpose [31]. Compositionally, SCMs must be rich in silica, alumina, and calcium, and the proportion of glassy to crystalline materials should be high to ensure that the material will be reactive in an alkaline environment [32]. The proportions of the glass oxides in SCMs will influence the reaction kinetics in a blended OPC/SCM system, as well as the thermodynamic stability of the reaction products [33].

3.2 Wind turbine GF composition and chemistry

The GF used in wind turbine composites is compositionally similar to that of traditional SCM (**Figure 5**); however, its amorphous nature may make it more reactive than

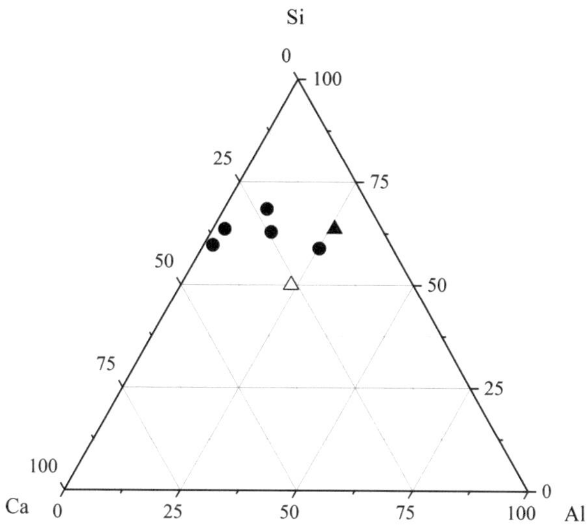

Figure 5.
Ternary diagram showing GF (closed circles) and class C (open triangle) and class F (closed triangle) fly ash from [8].

highly crystalline materials, suggesting that it may be more pozzolanic than many fly ashes. The GF is initially derived from geologic sources, which are melted in a furnace at high temperatures (approx. 1400°C) and passed through a perforated bushing before mechanical attenuation and fiberization [34, 35]. The resulting fibers exhibit varying compositions that depend on the initial proportions of raw materials, but are generally high in silica, calcium, and aluminum oxides (**Table 2** and **Figure 6**) [8].

Chemical compositions for GF used in general applications [16, 36–38] as well as wind-turbine-specific applications [19, 39] are characterized in literature (**Table 2** and **Figure 6**). Glosser et al. [8] condensed the compositional variation of GF into three notional end member compositions that are statistically representative of GF used in wind turbines and normalized to 100% to account for differences in reporting of LOI and resin between the derivative studies (**Table 2**). As seen in both **Figure 6 Table 2**, despite the variation in specific GF compositions, the primary chemical constituents of GF used in wind turbines – SiO_2, CaO, and Al_2O_3 – are congruent with those of OPC and SCMs like fly ash and slag. Section 4 discusses further assessment by Glosser et al. [8] regarding the potential of GF from wind turbines as an SCM, using thermodynamic modeling to predict important characteristics of cement + GF mixtures.

3.3 Review of thermodynamic modeling in cementitious systems

Thermodynamic modeling of cementitious systems is a proven technique to predict the type and amount of solid reaction products and pore solution chemistry, based only on a-priori knowledge of the chemistry of the reactants [24, 40–43]. Equilibrium calculations can be performed using a Gibbs Energy Minimization algorithm – which assumes that the reaction proceeds to infinite time [44]. This makes thermodynamic modeling an attractive alternative to experimental designs of OPC/SCM mixtures, particularly when the goal is to predict and optimize for a particular proportion of reaction products.

	HC (g/100 g)	MC/MS (g/100 g)	HS (g/100 g)
CaO	38.00 (34.63)	17.79 (28.95)	22.82 (26.06)
SiO_2	58.00 (52.86)	29.94 (48.73)	51.01 (58.27)
Al_2O_3	8.58 (10.64)	8.58 (13.96)	8.58 (9.81)
Fe_2O_3	3.50 (4.32)	3.50 (5.69)	3.50 (3.99)
Na_2O	0.54 (0.49)	0.54 (0.88)	0.54 (0.62)
MgO	0.56 (0.41)	0.56 (0.91)	0.56 (0.64)
SO_3	0.06 (0.05)	0.06 (0.10)	0.06 (0.07)
K_2O	0.20 (0.18)	0.20 (0.33)	0.20 (0.23)
Sr	0.18 (0.16)	0.18 (0.29)	0.18 (0.21)
Ba	0.03 (0.02)	0.03 (0.05)	0.03 (0.02)

Table 2.
Chemical composition of notional GF end members raw and (normalized to 100) from [8].

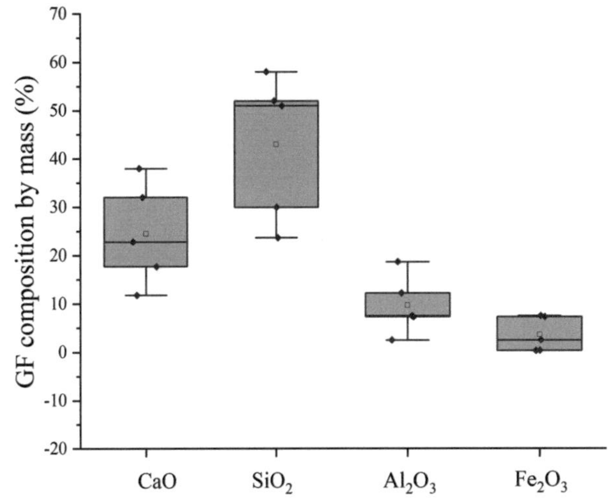

Figure 6.
GF composition as derived from literature. Figure from Glosser et al. [8].

4. Potential use of GF as SCM

4.1 Review of thermodynamic study of GF as an SCM

To date, only one known study has evaluated wind turbine GF for use as an SCM. In this study by Glosser et al. [8], thermodynamic simulations were performed to model the pore solution chemistry and major reaction products of cement + GF mixtures of varying GF end-member compositions derived from literature reports [16, 19, 38, 39]. Replacement levels of GF up to 60% by mass were evaluated and compared to results for OPC. Three notional GF end members were evaluated to capture a probable range of compositions: 1) high CaO (HC); 2) median CaO, median SiO_2 (MC/MS); and 3) high SiO_2 (HS). These end members (**Table 2**) are the CaO and SiO_2 values (g/100 g) of the minimum, median, and maximum CaO:SiO_2 values in grams/100 grams as reported in literature.. In all cases, the mean of Fe_2O_3 and Al_2O_3 (grams/100 grams) of all literature values were used. For the trace metals, the maximum of the reported mass concentrations were selected. The compositions of each notional GF composition were then normalized to 100%. Details of the methods can be found in the original study [8].

As discussed in Section 3 of this chapter, two main solid reaction products of interest in a cementitious system are CH and C-S-H. The study modeled the CH and C-S-H content of fully hydrated GF + OPC systems for all GF end members with replacement levels ranging from 0–60% by mass [8]. Results of this study found that for all replacement levels, mixtures with any GF end member contained less CH by mass than OPC only mixtures. This was found to be a result of the consumption of CH in the pozzolanic reaction, separate from dilutionary effects. The authors compared the GF + OPC CH results to that of a fly ash of identical bulk composition to the MC/MS GF, but 40% glass content, and found that the CH would not be fully consumed under these experimental conditions. The implications of this result suggest that GF is a strongly pozzolanic SCM. Examination of the C-S-H content of the GF mixtures supports this interpretation [8].

In keeping with the observations of CH, C-S-H amounts were found to be greater for all GF end members than in OPC, and showed varying trends based on the overall silica content of the GF (**Figure** 7). Recall from Eqs. (1) and (2) that C-S-H forms from the pozzolanic reaction between silica and CH, so it would be expected that greater concentrations of C-S-H would be present in mixtures where more CH is consumed by this reaction. As seen in **Figure** 7, this trend was found in the study, with C-S-H concentrations peaking for each end-member mixture at the same GF replacement level as the CH depletion. In all cases and at all replacement levels, the GF + OPC mixtures were found to contain more C-S-H than an OPC only mixture. The study further compared the GF + OPC mixtures to a fly ash with an identical bulk composition as the MC/MS end member, but 40% glass content, and found that the fly ash mixture produced less C-S-H than the GF + OPC mixture. [8]. Since C-S-H is generally considered to be a favorable hydrate, the implications of these simulations are clear: GF may be a powerful SCM for optimizing the amount of C-S-H in cementitious mixtures. The study further found that certain trace metals such as strontium reacted with the cement clinker to form insoluble strontium hydrates, which has the benefit of immobilizing potentially ecotoxic minerals from the GF.

The pore solution composition makes up the other important component of cement. The study also simulated pore solution pH and composition for each GF end member, up to replacement levels of 60%, and found that pore solution pH for each GF end member were similar up until roughly the 30% replacement level. After eclipsing the 30% replacement level the pH for all end members decreased, and eventually declined to values lower than that of OPC (pH = 13.6). These trends were found to correlate with the depletion of CH and the drop in alkali hydroxides in the pore solution. The results of these simulations show that cements made with GF may be used to achieve a range of optimal pH, depending on the mixture design specifications and the intended use of the mixture. The authors noted that the pore solution pH values of these mixtures, particularly at high replacement levels, is equivalent to those achieved with high reactivity fly ashes. This was found to be true despite the overall lower alkali content in the GF relative to fly ash. The authors stated that the similarity in results between GF and fly ash is likely due to the higher crystallinity and therefore lower reactivity of most fly ashes, which results in a lower available reactive alkali content than that the bulk concentrations of the ashes [8].

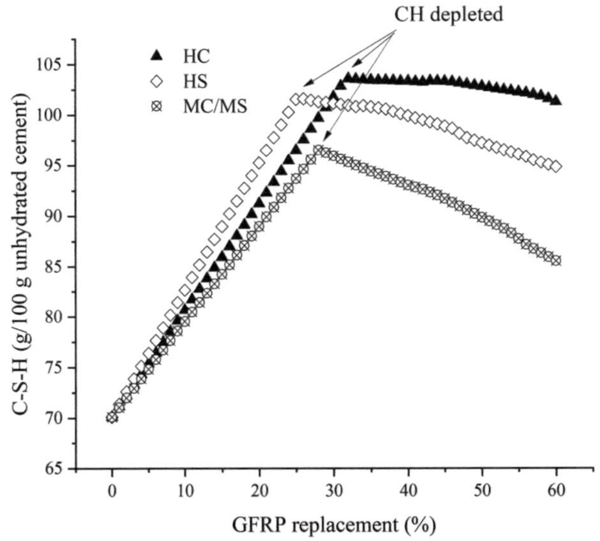

Figure 7.
C-S-H content of GF + OPC end members at replacement levels from 0–60%, with CH depletion point demarcated by arrow. Adapted from Glosser et al. [8].

4.2 Discussion of GF as an SCM

In this section, the potential for revalorization of wind turbine GF as an SCM is critically discussed in the context of the waste hierarchy shown in **Figure 3**, as well with an appreciation for the economic viability and practical benefits and drawbacks based on chemistry and intended use.

As shown in **Figure 3**, recycling of GFRP falls below repurposing and prevention as a preferred waste strategy, but above disposal. The revalorization of wind turbine GF as an SCM would be categorized as a recycling process, insofar as the GF is separated from the polymer (**Figure 2**) and turned into a new substance: in this case, a powder comprised of ground glass fibers. A major limitation of this use of GF therefore, is the energetic cost and processing required to separate the GF from the resin, and then grind it into a suitably fine powder for it to react with the cement clinker. It is worth noting that the carbon emissions from such a process will be greatly reduced in future years, as the transition away from fossil fuels to renewable energy progresses. This may reframe the position of recycling in the waste reuse hierarchy. This will be particularly true if the separated resin material can be beneficiated in an energy capture process, such as is currently done in the case of cement co-processing as described in Section 3. Technological improvements which increase the yields of GF from chemical and thermal separations will also be needed for recycling of the fibers to become viable at the scale needed for wind turbine blade waste.

Another current limitation of the revalorization of wind turbine GF as an SCM is economic: As described in Section 2 of this chapter, the most common end-of-life option for wind turbines is repurposing, due to the expense associated with separating GF from the resin, as well as the low yields and loss of mechanical strength of the fibers following separations. The issue of expense is likely the limiting factor at present time for revalorizing the GF from wind turbines as an SCM. However, the European co-processing projects described in Section 3 are indicative that the costs need not be prohibitive when the processes are scaled. This may be particularly true in coming decades, when the amount of wind turbine waste entering the waste stream renders repurposing of such a large volume of waste impractical and uneconomical. Ongoing research into commercializing and upscaling thermal and chemical separations processes described in Section 2 of this chapter will also alter the economic balance of recycling wind turbine waste for use as an SCM. As these separations processes become more efficient and more common, the associated cost will decrease, making them more attractive options.

While the limitations and drawbacks to recycling the GF extracted from wind turbine GFRP as an SCM are well articulated, it is worth noting that there are some discrete benefits for this use, both at present as well as in future scenarios. The first present benefit of revalorizing the GF following separation from the resin is that the mechanical properties of the fibers are irrelevant to its use as an SCM. As discussed in Section 2 of this chapter, following chemical and thermal separation of the fibers from the resin, the resulting glass fibers suffer a loss of mechanical strength. This presents a physical and engineering limitation on how the fibers can be reused, as the intended application must not rely on the mechanical strength of the fibers being preserved. However, when considering the use of recycled GF as an SCM, it is the chemical composition and amorphous nature of the fibers that makes them valuable, rather than their mechanical properties. Thus, the loss of mechanical strength is immaterial to their performance for this use case.

Another advantage to the beneficiation of glass fibers for use as an SCM is that the chemical composition of the feedstock need not meet any particular specification, so long calcium, silicon, and aluminum oxides make up the bulk of the glass. The need for a precise fiber chemistry has hampered the use of these fibers in the cement co-processing use case. However, as shown in the study by Glosser et al. [8], a wide range of GF chemistries are acceptable and perform well in cement mixtures as an SCM. Therefore, variations in the proportions of chemical oxides in the GF feedstock will not adversely impact performance. In fact, a variety of compositions may be beneficial, as this would enable GF + OPC mixture designs that could meet a range of desired specifications.

An additional advantage of using the GF from wind turbine GFRP as an SCM is the ability of the trace metals present in the fibers to react with the cement powder to form insoluble hydrates. As shown in the study by Glosser et al. [8], the presence of contaminants – which renders recycled GF unusable for certain applications where impurities are detrimental – not only does not negatively impact its performance as an SCM, but may be an effective method to manage ecotoxicities present in the fibers. Furthermore, the glassy nature of the fibers has been shown to render them more chemically reactive than current SCM feedstocks such as fly ash, which is of benefit for the design of cement mixtures.

Finally, the use of GF from wind turbine blade GFRP as an SCM has the benefit of offsetting CO_2 emissions from the cement clinkering process, which is a major environmental issue, as the cement industry accounts for 8% of emissions annually. When put in this context, the inefficiencies and energy expense inherent in current GF separations processes are partially offset.

5. Conclusion

This book chapter provides a review of the sources, processing, structure and chemistry of glass fiber materials used in wind turbines, its end-of-life options through both repurposing of whole glass fiber reinforced polymer parts and separations and recycling of glass fibers, and recapitulates the results of a study evaluating a novel use of the fibers as a supplementary cementitious material. The current and future benefits and limitations of such revalorization is presented within a critical structure that examines the scientific and engineering processes in Ref. to probable environmental impacts as well as economic constraints.

The physical structure of GFRP is examined in this chapter through both scanning electron microscopy as well as a review of literature. It is shown that GFRP composites are comprised of fibrous filaments made up of glass fibers, which are integrated into a thermoset polymeric resin. The fibers are derived from geologic sources that have been melted and processed, and are primarily comprised of silicon, calcium, and aluminum oxides.

This chapter has shown that GFRP blade waste from wind turbines presents a complex waste management challenge. This is due to in part to the large size of turbine blades, as well as the substantial amount of waste materials that will be entering the stream in coming decades as more wind turbines are retired. The thermoset nature of the GFRP, as well as the economic and energetic expenses associated with separating and refining the GF from the matrix also complicate its end-of-life uses.

A review of scientific literature described in this chapter demonstrates that at present, repurposing of wind turbine blades not only has the lowest environmental impact in terms of a waste management hierarchy, but also is the most common present solution in large part due to the low cost and substantial current commercial operations. Recycling of wind turbine blade GFRP is less common. Currently, the major limitations involved with separating and recycling GF components are low yield of glass fibers; high energetic expense of the processes; and the presence of impurities in the resulting products. However, the energetic and economic costs associated with recycling technologies are expected to be reduced in coming decades, as separations processes become more efficient and economical with scale.

The use of recycled GF from wind turbine blades is evaluated in the context of the cement industry, with a specific focus on its use as a supplementary cementitious material. Compositionally, the chemical constituents of glass fibers are shown to be nearly identical to that of cement clinker, and therefore, show promise for its use in cement mixtures. A thermodynamic study which models the reaction products of simulated glass fiber and cement mixtures is recapitulated and critically evaluated. Study results suggest that the chemistry and reactivity of glass fibers may make them suitable for this use, and that they are expected to react with cement clinker to produce pore solution compositions and solid reaction products that are beneficial for cement strength and durability. The major benefits of this use include the high chemical reactivities of the fibers; the reaction of impurities in the recycled fibers with cement constituents to immobilize contaminants; and the range of acceptable GF feedstock compositions that can form favorable cement reaction products. However, it is shown that the current cost and lack of commercial operations for separating GF from resin makes such revalorization uneconomical at present. In coming decades however, the increase in wind turbine waste may render this option both more practical as well as more economical. Furthermore, for a large and complex material like wind turbines, recycling may become the most attractive option, as repurposing the parts will become impractical given the large volume of waste expected in coming decades.

Acknowledgements

The authors wish to acknowledge Prannoy Suraneni, Eli Santkyul, and Eric Fagan who co-authored the original study reported in this chapter. We also wish to thank Dr. Michael Kraft of WWU for his expertise with the scanning electron microscope and Dr. Mark Peyron for the GFRP sample.

Conflict of interest

The authors declare no conflict of interest.

Author details

Deborah Glosser[1,2,3]*, Layla Russell[1] and Paul Striby[1]

1 Institute for Energy Studies, Western Washington University, Bellingham WA, USA

2 Engineering and Design, Western Washington University, Bellingham, WA, USA

3 Advanced Materials Science and Engineering Center, Bellingham, WA, USA

*Address all correspondence to: glossed@wwu.edu

References

[1] Dickson AN, Barry JN, McDonnell KA, et al. Fabrication of continuous carbon, glass and Kevlar fibre reinforced polymer composites using additive manufacturing. Additive Manufacturing. 2017;**2017**:16

[2] Fouad Y, El-Meniawi M, Afifi A. Erosion behaviour of epoxy based unidirectionl (GFRP) composite materials. Alexandria Engineering Journal. 2011;**50**:29-34

[3] Wu SH, Wang FY, Ma CCM, et al. Mechanical, thermal and morphological properties of glass fiber and carbon fiber reinforced polyamide-6 and polyamide-6/clay nanocomposites. Materials Letters. 2001;**49**:327-333

[4] Gamze Karsli N, Yilmaz T, Aytac A, et al. Investigation of erosive wear behavior and physical properties of SGF and/or calcite reinforced ABS/PA6 composites. Composite B Engineering. 2013;**44**:385-393

[5] Chen S, Yang Q, Brow RK, et al. In vitro stimulation of vascular endothelial growth factor by borate-based glass fibers under dynamic flow conditions. Materials Science and Engineering: C. 2017;**73**:447-455

[6] Sathishkumar TP, Satheeshkumar S, Naveen J. Glass fiber-reinforced polymer composites-a review. DOI: 10.1177/0731684414530790

[7] Dang W, Kubouchi M, Maruyama T, et al. Decomposition of GFRP in nitric acid and hydrogen peroxide solution for chemical recycling. Progress in Rubber, Plastics and Recycling Technology. 2002;**18**:49-67

[8] Glosser D, Santykul E, Fagan E, et al. A thermodynamic perspective on wind turbine glass fiber waste as a supplementary cementitious material. Cement. 2022;**9**:100039

[9] Liu P, Barlow CY. Wind turbine blade waste in 2050. Waste Management. 2017;**62**:229-240

[10] Wind Turbines: The Bigger, the Better | Department of Energy, https://www.energy.gov/eere/articles/wind-turbines-bigger-better. (Accessed 4 October 2022)

[11] Wei Y, Hadigheh SA. Cost benefit and life cycle analysis of CFRP and GFRP waste treatment methods. Construction and Building Materials. 2022;**348**:128654

[12] Gentry TR, Al-Haddad T, Bank LC, et al. Structural analysis of a roof extracted from a wind turbine blade. Journal of Architectural Engineering. 2020;**26**:04020040

[13] Rubert T, Niewczas P, Mcmillan D. Life Extension for Wind Turbine Structures and Foundations.

[14] Bank LC, Arias FR, Yazdanbakhsh A, et al. Concepts for reusing composite materials from decommissioned wind turbine blades in affordable housing. Recycling. 2018;**2018**:3

[15] Karuppannan Gopalraj S, Kärki T. A review on the recycling of waste carbon fibre/glass fibre-reinforced composites: Fibre recovery, properties and life-cycle analysis. Epub ahead of print 123AD. DOI: 10.1007/s42452-020-2195-4.

[16] Asokan P, Osmani M, Price ADF. Assessing the recycling potential of glass fibre reinforced plastic waste in concrete and cement composites. Journal of Cleaner Production. 2009;**17**:821-829

[17] Zhou B, Zhang M, Wang L, et al. Experimental study on mechanical property and microstructure of cement mortar reinforced with elaborately recycled GFRP fiber. Cement and Concrete Composites. 2021;**117**:103908

[18] Decommissioned wind turbine blades used for cement co-processing | CompositesWorld, https://www.compositesworld.com/articles/recycled-composites-from-wind-turbine-blades-used-for-cement-co-processing. (Accessed 21 October 2022)

[19] Nagle AJ, Delaney EL, Bank LC, et al. A comparative life cycle assessment between landfilling and co-processing of waste from decommissioned Irish wind turbine blades. Journal of Cleaner Production. 2020;**277**:123321

[20] Pickering SJ. Recycling technologies for thermoset composite materials-current status. Composites. Part A, Applied Science and Manufacturing. 2006;**37**:1206-1215

[21] Ginder RS, Ozcan S. Recycling of commercial E-glass reinforced thermoset composites via two temperature step pyrolysis to improve recovered fiber tensile strength and failure strain. Recycling. 2019;**4**:24

[22] Kanerva M, Jokinen J, Sarlin E, et al. Lower stiffness of GFRP after sulfuric acid-solution aging is due to degradation of fibre-matrix interfaces? Composite Structures. 2019;**212**:524-534

[23] Lothenbach B, Scrivener KL, Hooton D, et al. Supplementary cementitious materials supplementary cementitious materials. Cement and Concrete Research. 2011;**41**:1244-1256

[24] Lothenbach B, Winnefeld F. Thermodynamic modelling of the hydration of Portland cement. Cement and Concrete Research. 2006;**36**:209-226

[25] de Weerdt K, Haha M ben, le Saout G, et al. Hydration mechanisms of ternary Portland cements containing limestone powder and fly ash. Cement and Concrete Research 2011; 41: 279-291.

[26] Glosser D, Azad VJ, Suraneni P, et al. Extension of powers-Brownyard model to pastes containing supplementary cementitious materials. ACI Materials Journal. 2019;**116**:205-216

[27] Nonat A. The structure and stoichiometry of C-S-H. Cement and Concrete Research. 2004;**34**:1521-1528

[28] Hong SY, Glasser FP. Alkali binding in cement pastes: Part I. The C-S-H phase. Cement Concrete Research. 1999;**29**:1893-1903

[29] Isgor OB, Razaqpur a G. Modelling steel corrosion in concrete structures. Materials and Structures. 2006;**53**:291-302

[30] Bharadwaj K, Isgor OB, Weiss WJ, et al. A new mixture proportioning method for performance- based concrete. Materials Journal. 2022;**119**:207-220

[31] Glosser D, Jafari Azad V, Isgor B, et al. An extension of the powers-Brownyard model to pastes containing SCM. ACI Materials Journal.

[32] Glosser D, Suraneni P, Isgor OB, et al. Using glass content to determine the reactivity of fly ash for thermodynamic calculations. Cement Concrete Composite. 2021;**2021**:115

[33] Glosser D, Burkan Isgor O, Jason WW. Non-equilibrium thermodynamic modeling framework for ordinary Portland cement/ supplementary cementitious material systems. ACI Materials Journal. 2020;**117**:111-123

[34] GFRP: Manufacturing process, properties and application, https://www.pfh-university.com/blog/gfrp-manufacturing-process-properties-and-application.html. (Accessed 21 October 2022)

[35] Stickel JM, Nagarajan M. Glass Fiber-reinforced composites: From formulation to application. International Journal of Applied Glass Science. 2012;**3**:122-136

[36] Dehghan A, Peterson K, Shvarzman A. Recycled glass fiber reinforced polymer additions to Portland cement concrete. Construction and Building Materials. 2017;**146**:238-250

[37] Sathishkumar TP, Satheeshkumar S, Naveen J. Glass fiber-reinforced polymer composites - A review. Journal of Reinforced Plastics and Composites. 2014;**33**:1258-1275

[38] Correia JR, Almeida NM, Figueira JR. Recycling of FRP composites: Reusing fine GFRP waste in concrete mixtures. Journal of Cleaner Production. 2011;**19**:1745-1753

[39] Novais RM, Carvalheiras J, Capela MN, et al. Incorporation of glass fibre fabrics waste into geopolymer matrices: An eco-friendly solution for off-cuts coming from wind turbine blade production. Construction and Building Materials. 2018;**187**:876-883

[40] Lothenbach B. Thermodynamic equilibrium calculations in cementitious systems. Materials and Structures/Materiaux et Constructions. 2010;**43**:1413-1433

[41] Lothenbach B, Kulik DA, Matschei T, et al. Cemdata18: A chemical thermodynamic database for hydrated Portland cements and alkali-activated materials. Cement and Concrete Research. 2019;**115**:472-506

[42] Lothenbach B, Matschei T, Möschner G, et al. Thermodynamic modelling of the effect of temperature on the hydration and porosity of Portland cement. Cement and Concrete Research. 2008;**38**:1-18

[43] Miron GD, Wagner T, Kulik DA, et al. An internally consistent thermodynamic dataset for aqueous species in the system Ca-Mg-Na-K-Al-Si-O-H-C-Cl to 800°C and 5 KBAR. American Journal of Science. 2017;**317**:755-806

[44] Kulik DA, Wagner T, Dmytrieva S v., et al. GEM-Selektor geochemical modeling package: Revised algorithm and GEMS3K numerical kernel for coupled simulation codes. Computational Geosciences 2013; 17: 1-24.

Chapter 2

Spectral Characterization and Analysis of Underground Optical Fibre Cable Network Using Optical Time Domain Reflectometry

Asiya E. Asiya, Michael U. Onuu, Rufus C. Okoro and O. Enendu Uche

Abstract

Many of the optical fibre cables comprised of 1310 nm zero-dispersion single-mode (SM) optical fibres installed in underground/conduits and access networks. Currently, there have been several studies on active network systems, which are designed to increase transmission capacity and flexibility. The application of active communication devices like the wavelength division multiplexing (WDM) systems, usually using SM optical fibre for transmission in the 1310–1625 nm window wavelength, proves very effective in decreasing the installation costs and high signal attenuations. It was imperative to examine the wavelength dependency of such transmission characteristics of SM optical fibre cables previously installed and in which several optical fibres were spliced. Analysis for such network has been performed and monitored over 1550–1625 nm wavelength. Results show that the spectral characterization and analysis of a long-haul optical network system operating at the 50-GHz-spaced 80-dense wavelength division multiplexing (DWDM)-channel can be used to identify the presence of faults.

Keywords: optical fibre cable, attenuation, chromatic dispersion, polarization mode dispersion, OSNR, WDM

1. Introduction

An optical network is a communications network in which the transmission links are made up of optical fibres and whose architecture aims at exploring optical fibre advantages such as high speed, higher bandwidth, greater reliability and lower maintenance cost [1].

The deployment of optical fibre cable as a backbone for the transmission of data is becoming more and more prevalent in the field of telecommunications. Telephone and cable companies and internet service providers, in collaboration with government authorities, for example, are continuously expanding their fibre networks to reach more consumers across a wider geographic area. Even though it is unlikely that there will be a need for gigabit speeds for desktop and home appliances (IoTs) any time soon, many applications are outgoing their 10 and 100 Mbps ethernet LANs [2].

Currently, most Chinese information capacity is transferred over the optic cable line. With the increase of optic fibre cable faults and optical fibre cable ageing, the frequency of optic fibre line faults increases, and it is difficult to detect fault's location in the traditional optical fibre management mode. It will take a long time to eliminate the failure factor that affects the regular work of the network [3]. Although nowadays ring network protection technology can guarantee smooth and continuous transmission to a certain extent, the shortcoming of traditional line maintenance still exists. Therefore, the implementation of the optic fibre cable line real-time detection and management, dynamic observations of the transmission properties of the optic fibre cable line degradation and the timely discovery may prevent hidden trouble and reduce the incidence of signal delaying. However, such management systems may only inform the maintainers when a fault happens but cannot detect the exact location of the fault.

During installation, cables or connectors may be broken, and other impacts such as bilging, stress and ageing on the cable and active devices can lead to faults in the entire transmission system. Such failures may result in economic loss and also causes great inconvenience to users' social well-being [4]. Therefore, it is essential to ensure the wellness of fibre optic cable. Repair and maintenance of cable communication have profound significance. The practical approach to faults determination focuses on measuring the faults distance of optical cables and the Euclidean distance to the earth's surface [5] between the optical transmitter and the point in the underground cable cut. However, the solution proposed in Ref. [6] introduced several cable junctions along the fibre optic cable transmission line, which ultimately increased the loss in the underground fibre optic network [7–9].

In engineering new optical systems on an older fibre link, it is frequently necessary to repair existing faults and other links between fibre optic cables. When technicians are called to the field to repair or modify fibre faults, they rely on manually produced maps to identify the proper location to dig for the cables. Because most underground optical fibre cable is buried adjacent to railroad tracks, digging becomes very expensive due to the extensive governmental procedures like giving right of way and, on the other hand, contract crews necessary to flag the railroad and traffic. A failed attempt to locate a fibre optic cable thus results in increased costs due to longer service times, the cost of restoring the improperly excavated area and the opportunity to damage other underground equipment that may be buried where the fibre optic cables were thought to be located.

In this paper, the authors examine the wavelength dependency of long-haul underground SM fibre optic cable networks and their corresponding active components with the view to characterize the spectral dependency of such communication network systems. It was possible to detect and trace faults with minimal error on the optical fibre cable network using the principle of the backscattering method of the optical time-domain reflectometer.

2. Literature review

According to work done in Refs. [5, 10], simulation and laboratory work provide relevant expected results in determining faults in the network. Various tools were used to conduct a series of laboratory, field and classroom experiments to achieve the accuracy of vulnerability tracing. These tools use different scientific principles, tools and measurement techniques to measure distances during troubleshooting. Previous studies on these scientific principles and other techniques include flaw detection using OTDR [11], a photonic probe flaw locator, Raman fibre sensor, T-OTDR, correlation technique using traffic signal and step frequency method [12]. Reviews of articles on fault detection focus mainly on fibre optic cables, except in Ref. [13], which provide theoretical results of FOC to fault distance measurement on the earth's surface. The mean time to repair (MTTR) of fault tracing in underground optical networks is high due to the complex processes involved. The distance from faults in the underground fibre optic network must be accurately and timely determined and correctly connected to restore interrupted services and maintain the region's customer experience. This can only be achieved if an intelligent system is integrated into the tracing technique for the prediction of fault distances on earth [14–17]. In this regard, the author in Ref. [18] has proposed an artificial intelligence model to complement the functions of conventional OTDRs.

2.1 Spectral characterization of optical fibre

The spectral characteristics of an optical fibre network may be categorized as linear or non-linear. The non-linear characteristics are subjective to some factors, like bit rates, channel spacing and power levels [19]. While the linear characteristics include attenuation, chromatic dispersion (CD), polarization mode dispersion (PMD) and optical signal-to-noise ratio (OSNR) [20, 21].

2.2 Attenuation

Fibre attenuation can be described by the expression given in Eq. (1):

$$^{dP}/_{dz} = -\alpha P \tag{1}$$

where α corresponds to the power attenuation factor per unit length.

If launched power into the fibre cable is P_{in}, then the output power after propagation via the fibre length, L, is given by Eq. (2):

$$P_{out} = P_{in}\, exp^{(-\alpha L)} \tag{2}$$

The absorption coefficient, α, varies with wavelength, λ. This coefficient characterizes the loss measured in decibels per kilometre length of the fibre [22].

Linear characteristic is either initiated by internal or external factors. The internal factor results from impure substances that are inherently present in the fibre during the manufacturing process. When light signals hit any of the impurities, it scatters or is absorbed, leading to an intrinsic loss. This loss can further be classified into two components:

i. The material absorption is caused by imperfection as well as impurities in the fibre.

ii. Rayleigh scattering results from the elastic interactions between the light wave and the glass molecules. It accounts for nearly 96% of attenuation in the optical fibre cable and varies as expressed in Eq. (3):

$$\alpha_R = {}^{c}/_{\lambda^4}, \text{where } C = 0.7 - 0.9 \ ({}^{dB}/_{km}) - \mu m^4 \tag{3}$$

On the other hand, extrinsic attenuations are initiated by external stains such as macro or micro bending. Imperfection in the cylindrical geometry of fibre during the manufacturing process leads to a macro bending and can be visible when inspected with the OTDR [6, 23]. In essence, when the bends are corrected, the loss generally is reversible. If loss must be avoided, the minimum bending radius must not exceed ten times the outer diameter of the cable type. Similarly, temperature, tensile stress or severe force may also affect micro-bending [24]. It is irreversible. Both bends will result in the reduction of optical output power.

2.3 OSNR

The IEC standard has defined OSNR as the signal power at the peak of a channel divided by the noise power interpolated at the position of the peak. This definition can be expressed in the form of Eq. (4):

$$OSNR = 10 \ log_{10} \left({}^{P_i}/_{N_i} \right) + 10 \ log_{10} \left({}^{B_m}/_{B_r} \right) \tag{4}$$

where P_i is the optical signal power at the i^{th}channel,
B_m is the resolution bandwidth,
N_i is the interpolated value of noise power and
B_r is the reference optical bandwidth.

An additional noise can also be added to the entire system by some devices like optical amplifiers, ROADMs and lasers. Thus, in this work, optical amplifier noise has been considered the main source of OSNR limitation and degradation.

2.4 Chromatic dispersion (CD)

Chromatic dispersion in fibre network links is a result of the change in group delay per unit wavelength in ps/nm. It accumulates with distance. The delay coefficient is generally quoted in units of ps/(nm*km) and depends on the fibre type (in this case, G.652).

The graph of group delay versus wavelength was then fitted to data using the approximation equation of (6) in **Table 1**.

2.5 Polarization mode dispersion (PMD)

PMD is measured in pico-seconds (ps) for a span length of installed fibre cables.

The general result of this phenomenon may be a continuous increment in PMD, as given in Eq. (5), the appropriate units that are utilized for the coefficient that characterizes the fibre itself are $ps/km^{½}$.

$$PMD_{total} = \left(\sum_n (PMD_n)^2 \right)^{½} \tag{5}$$

The polarization unit vector, representing the state of polarization (SOP) of the electric field vector, does not remain constant in practical optical fibres; rather, it changes in a random fashion along with the fibre because of its fluctuating

Fit technique	Group delay's equation	Dispersion data's equation	Equation no.
3-term Sellmeier	$A + B\lambda^2 + C\lambda^{-2}$	$2B\lambda + C\lambda^{-3}$	(6)
5-term Sellmeier	$A + B\lambda^2 + C\lambda^{-2} + D\lambda^4 + E\lambda^{-4}$	$2B\lambda^2 - 2C\lambda^{-3} + 4D\lambda^3 - 4E\lambda^{-5}$	(7)
2nd order polynomial (quadratic)	$A + B\lambda + C\lambda^2$	$B + 2C\lambda$	(8)
3rd-order polynomial (cubic)	$A + B\lambda + C\lambda^2 + D\lambda^3$	$B + 2C\lambda + 3C\lambda^2$	(9)
4th order polynomial	$A + B\lambda + C\lambda^2 + D\lambda^3 + E\lambda^4$	$B + 2C\lambda + 3C\lambda^2 + 4E\lambda^3$	(10)
Equations for slope (table A2/ITU-T G650.1)			
Fit technique	Dispersion slope's equation		
3-term Sellmeier	$2B + 6C\lambda^{-4}$		(11)
5-term Sellmeier	$2B + 6C\lambda^{-4} + 12D\lambda^2 + 20E\lambda^{-6}$		(12)
2nd order polynomial (quadratic)	$2C$		(13)
3rd-order polynomial (cubic)	$2C + 6D\lambda$		(14)
4th order polynomial	$2C + 6D\lambda + 12D\lambda^2$		(15)
Equations for zero dispersion wavelength and slope (A.3/ITU G650.1)			
Fit technique	Zero dispersion wavelength	Zero dispersion slope	
3-term Sellmeier	$(C/B)^{1/4}$	$8B$	(16)
2nd order polynomial (quadratic)	$-B/(2C)$	$2C$	(17)

Table 1.
Approximation equations for group delay and dispersion [25].

birefringence. Two common birefringence sources are (1) geometric birefringence (related to small departures from perfect cylindrical symmetry) and (2) anisotropic stress (produced on the fibre core during manufacturing or cabling of the fibre).

The degree of birefringence is described by the difference in refractive indices of orthogonally polarized modes of Eq. (18).

$$B_m = \left|n_x - n_y\right| = \Delta n \tag{18}$$

According to Ref. [26], the corresponding difference in propagation constants of two orthogonally polarized modes is given by Eq. (19).

$$\Delta\beta = \left|\beta_x - \beta_y\right| = (\omega/c)\Delta n \tag{19}$$

Birefringence causes the two polarization components to exchange power periodically, as described by the beat length of Eq. (20).

$$L_B = \frac{2\pi}{\Delta\beta} = \lambda/\Delta n \tag{20}$$

Typically, $B_m. \sim 10^{-7}$, and therefore $L_B \sim$10 m for $\lambda \sim$1 μm. Linearly polarized light remains linearly polarized only when it is polarized along with one of the principal axes; otherwise, its polarization state changes along the fibre length from linear to elliptic, then returns to linear, in a periodic manner, over the length L_B

The modal group indices and modal group velocities are related by Eq. (21).

$$n_{gx,y} = \frac{c}{v_{gx,y}} = \frac{c}{\left(1/\beta_{1x,y}\right)} \tag{21}$$

where

$$\beta_{1x,y} = \frac{d\beta_{x,y}}{d\omega} \tag{22}$$

So that the difference in time arrivals (at the end of fibre of length, L) for two orthogonal polarization modes, known as the differential group delay (DGD), can be calculated using Eq. (23).

$$\Delta\tau = \left|\frac{L}{v_{gx}} - \frac{L}{v_{gy}}\right| = L\left|\beta_{1x} - \beta_{1y}\right| = L\Delta\beta_1 \tag{23}$$

3. Methods

The optical time domain reflectometry technique has been employed in CD measurement. During the process, the OTDR sends pulsations of four or more wavelengths into each of the tested fibre cores so that the comparative arrival time is subsequently measured and recorded for each backscattered wavelength signal [27–29]. The comparison between the reference wavelength and the arrival times of the other wavelengths were computed and fitted to data using the approximation equations of (5) in **Table 1** above.

Similarly, a PMD test set was used to measure PMD. An averaging procedure helps in determining the PMD of several sections of the optical fibre link. The optical spectrum analyzer, OSA, has been deployed to automatically calculate the total PMD of the several spans in the network. This device uses the root mean square summation [30] of Eq. (8).

To characterize the entire network components, the OSA became the protagonist for such tests and measurements at two distinct ports on the WDM system using different wavelengths [31, 32]. The basic measurements in the frequency domain required were:

a. Total power for the optical signal

b. Measurement of channel power

c. OSNR

d. Measurement of central channel wavelength

e. Measurement of the spacing between signals [33]

The measurements were carried out with the FTB 5240S and FTB-5250S-EI instruments, designed with dedicated algorithms for each application.

The characterization measurements of the network [34] components, such as the ROADMs or filters, amplifiers, DFB lasers, FB laser and LED, were determined by the equipment in all scopes of application.

4. Results and discussion

4.1 Experimental process

Two experimental procedures were carried out to ascertain the presence of optical faults on the cable. These were as follows:

4.1.1 Chromatic dispersion (CD)

The obtained CD result is tabulated in **Table 2**. From this experiment, it was observed that the maximum allowable dispersion of 110.682 $ps/nm * km$ appears at 1548.50 nm wavelengths. Beyond this wavelength, signals failed to be transmitted due to the presence of CD caused by the 1550 nm modulator.

The relationship between the dispersion and the relative group delay (RGD) to wavelength has been presented in **Figure 1** for the 80 km SM fibre cable link. The graph reveals the presence of chromatic dispersion at the targeted wavelength of 1550 nm of the modulator; the value of CD is seen to be 16.696 $ps/(nm * km)$ with a slope of 0.060859 $ps/(nm2 * km)$. The RGD-weighted RMS error was 286.313 ps during the 4 seconds average acquisition time over the 0.5 nm wavelength step.

4.1.2 Polarization mode dispersion (PMD)

As depicted in **Figure 2**, the resulting value of PMD for a 40 km fibre length tested was 0.785 ps within the wavelength band of 1514.62–1588.66 nm. The PMD coefficient was 0.1241 $ps/k^{1/2}$ with a Gaussian compliance factor of 1.069.

The PDM and fibre loss over different lengths of optical fibre cable are presented in **Table 3**.

This depicts that both the link polarization dispersion and link losses increased with an increase in cable length.

Table 4 depicts the maximum PMD over a given transmission bit rate. The maximum PMD decreases with an increase in bit rate.

4.2 Drift analysis with WDM investigator

Figure 3 shows how power, wavelengths and OSNR were monitored over time in the SMFOC network. This was achieved using FTBx-5245/5255 OSA with a WDM investigator. The investigation indicates the presence and strength of polarization pulse spreading (PPS) in the tested channels. The controlled emissions of a 50-GHz-spaced 40-DWDM-channel covering the wavelength of 1529.545–1556.54 nm window have been presented. The investigator gave details on the fibre link characteristics. For example, the presence of carved noise from the PolMux and other types of impairments like carrier leakages, PMD pulse spreading, crosstalk and non-linear effects were also revealed by a certain degree of severity ('OK', 'warning', 'risk') as shown in

Used	Wavelength (nm)	RGD (ps)	Fitted RGD (ps)	RGD deviation (ps)	Pass/ Fail	Dispersion ps/nm	Pass/ fail	Dispersion ps/(nm*km)
Yes	1530.0	0.00	35.472	35.472	Pass	123.519	Pass	103.043
Yes	1530.5	628.78	654.360	25.579	Pass	1239.033	Pass	103.253
Yes	1531.0	1222.47	1274.505	52.039	Pass	1241.544	Pass	103.462
Yes	1531.5	1857.76	1895.904	38.145	Pass	1244.054	Pass	103.671
Yes	1532.0	2488.23	2518.558	30.333	Pass	1246.562	Pass	103.880
Yes	1532.5	3096.88	3142.466	45.583	Pass	1249.067	Pass	104.089
Yes	1533.0	3741.69	3767.626	25.938	Pass	1251.571	Pass	104.298
Yes	1533.5	4388.24	4394.036	5.798	Pass	1254.072	Pass	104.506
Yes	1534.0	4995.16	5021.698	26.537	Pass	1236.572	Pass	104.714
Yes	1534.5	5634.26	5650.608	16.345	Pass	1259.069	Pass	104.922
Yes	1535.0	6275.14	6280.766	5.630	Pass	1261.565	Pass	105.130
Yes	1535.5	6877.73	6912.172	32.447	Pass	1264.058	Pass	105.338
Yes	1536.0	7533.12	7544.824	11.701	Pass	1266.549	Pass	105.546
Yes	1536.5	8187.90	8178.721	9.184	Pass	1269.038	Pass	105.753
Yes	1537.0	8794.54	8813.862	19.319	Pass	1271.526	Pass	105.960
Yes	1537.5	9453.78	9450.246	3.535	Pass	1274.011	Pass	106.168
Yes	1538.0	10093.42	10087.873	5.548	Pass	1274.494	Pass	106.375
Yes	1538.5	10712.04	10726.740	14.698	Pass	1278.975	Pass	106.581
Yes	1539.0	11358.12	11366.848	8.725	Pass	1281.454	Pass	106.788
Yes	1539.5	12006.86	12008.194	1.337	Pass	1283.932	Pass	106.994
Yes	1540.0	12616.22	12650.779	34.556	Pass	1286.407	Pass	107.201
Yes	1540.5	13286.16	13294.601	8.445	Pass	1288.880	Pass	107.407
Yes	1541.0	13947.51	13939.658	7.848	Pass	1291.351	Pass	107.613
Yes	1541.5	14577.69	14585.951	8.262	Pass	1293.820	Pass	107.818
Yes	1542.0	15245.94	15833.478	12.461	Pass	1296.287	Pass	108.024
Yes	1542.5	15901.85	15882.238	19.613	Pass	1298.752	Pass	108.229
Yes	1543.0	16523.19	16532.230	9.040	Pass	1301.216	Pass	108.435
Yes	1543.5	17195.07	17183.454	11.617	Pass	1303.677	Pass	108.640
Yes	1544.0	17856.00	17835.907	20.089	Pass	1306.136	Pass	108.845
Yes	1544.5	18486.43	18489.589	3.155	Pass	1308.593	Pass	109.049
Yes	1545.0	19158.46	19144.500	13.959	Pass	1311.049	Pass	109.254
Yes	1545.5	19822.68	19800.638	22.043	Pass	1313.502	Pass	109.458
Yes	1546.0	20453.18	20458.001	4.822	Pass	1315.953	Pass	109.663
Yes	1546.5	21122.60	21116.590	6.012	Pass	1318.403	Pass	109.867
Yes	1547.0	21787.91	21776.404	11.506	Pass	1320.850	Pass	110.071
Yes	1547.5	22414.29	22437.440	23.147	Fail	1323.295	Pass	110.275
Yes	1548.0	23098.45	23099.699	1.246	Fail	1325.739	Pass	110.478

Used	Wavelength (nm)	RGD (ps)	Fitted RGD (ps)	RGD deviation (ps)	Pass/ Fail	Dispersion ps/nm	Pass/ fail	Dispersion ps/(nm*km)
Yes	1548.5	23785.11	23763.179	21.934	Fail	1328.180	Pass	110.682
Yes	1549.0	22409.30	24427.879	18.578	Fail	1330.620	Fail	110.885
Yes	1549.5	25100.09	25093.798	6.288	Fail	1333.058	Fail	111.088
Yes	1550.0	25770.91	25760.936	9.978	Fail	1335.493	Fail	111.291

Table 2.
Group delay and chromatic dispersion.

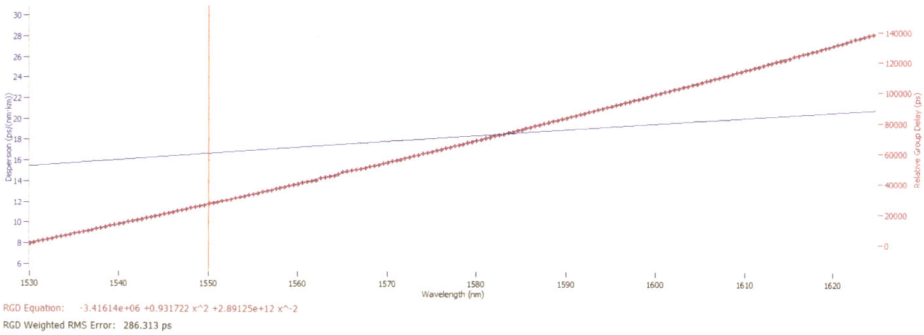

Figure 1.
Chromatic dispersion of 80 km fibre link.

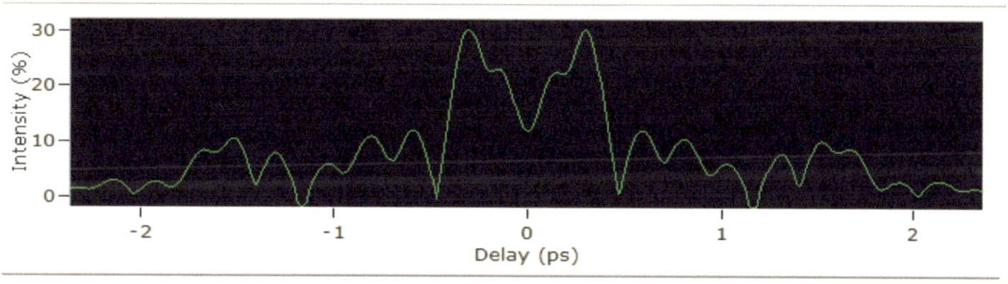

Figure 2.
PMD measured with the interferometry method.

Table 5 below. In this way, the status of the channels tested in a single port was visualized as a trace for any acquisition and change of state.

On the other hand, **Table 6** depicts the channel results for the signal power, OSNR, noise and bandwidth at 3.00 and 20.00 dB, respectively.

It is observed that channel C_025 has the best signal power with a minimum OSNR. Thus, multiple DWDM channels' output power was successfully controlled, and the target level functions were also achieved for the 1550 nm wavelength, as shown in **Figure 4**.

4.3 EDFA analysis

The EDFA was tested and recorded as depicted in **Table 7a** and **b**. OSA traced reports for both the input and output signals of the EDFA under consideration are

Link length (km)	Link PMD (ps)	Link loss (hr/ps)
40	0.785	0.41
200	1.43	0.38
400	2.42	0.18
600	2.53	0.14
800	2.80	0.09
1000	3.09	0.075
1,200	3.54	0.072
1,400	3.68	0.59
1,600	3.99	0.59

Table 3.
Comparison of link PMD and link loss over fibre link length [35].

Bite rate (Gbps)	Maximum PMD (ps)	PMD coefficient ((ps/km$^{1/2}$)
2.5	40	2.0
10	10	0.5
20	5	0.25
40	2.5	0.125

Table 4.
Maximum PMD value for a given bit rate used [35].

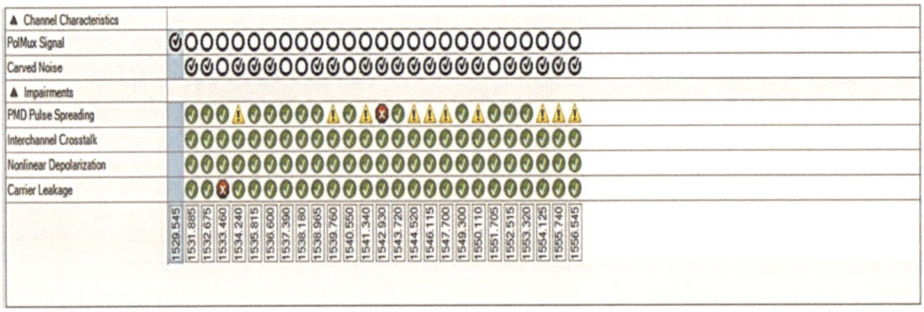

Figure 3.
WDM system with WDM investigator.

depicted in **Figure 5a** and **b**, respectively. From these results, it is observed that better performance of the EDFA has an average wavelength of 1550.42 nm with a total input power of 12.68 *dBm* and power flatness of 3.11 *dB* at the input channel. These parameters' values were, however, depreciated to an average power output of −6.21 *dBm* with a total power of −0.19 *dB* and a power flatness of 2.93 *dB* at the output channel. This is due to additional noise introduced by different active and passive components on the link.

Ch. #	Name	PolMux signal	Carved noise	PMD pulse spreading	Inter-channel crosstalk	Non-linear depolarization	Carrier leakage
1	C_001	Present					
2	C_002	'Not present'	'Present'	'OK'	'OK'	'OK'	'OK'
3	C_003	'Not present'	'Present'	'OK'	'OK'	'OK'	'OK'
4	C_004	'Not present'	Not present	'OK'	OK	OK	'Risk'
5	C_005	'Not present'	'Present'	'Warning'	'OK'	'OK'	'OK'
6	C_006	'Not present'	'Present'	'OK'	'OK'	'OK'	'OK'
7	C_007	'Not present'	'Present'	'OK'	'OK'	'OK'	'OK'
8	C_008	'Not present'	'Not present'	'OK'	'OK'	'OK'	'OK'
9	C_009	'Not present'	'Not present'	'OK'	'OK'	'OK'	'OK'
10	C_010	'Not present'	Present	'OK'	'OK'	'OK'	'OK'
11	C_011	'Not present'	Present	'Warning'	'OK'	'OK'	'OK'
12	C_012	'Not present'	'Not present'	'OK'	'OK'	'OK'	'OK'
13	C_013	'Not present'	Present	'Warning'	'OK'	'OK'	'OK'
14	C_014	'Not present'	'Present'	'Risk'	'OK'	'OK'	'OK'
15	C_015	'Not present'	'Present'	'OK'	'OK'	'OK'	'OK'
16	C_016	'Not present'	'Present'	'Warning'	'OK'	'OK'	'OK'
17	C_017	'Not present'	'Present'	'Warning'	'OK'	'OK'	'OK'
18	C_018	'Not present'	'Present'	'Warning'	'OK'	'OK'	'OK'
19	C_019	'Not present'	'Present'	'OK'	'OK'	'OK'	'OK'
20	C_020	'Not present'	'Present'	'Warning'	'OK'	'OK'	'OK'
21	C_021	'Not present'	'Not present'	'OK'	'OK'	'OK'	'OK'
22	C_022	'Not present'	'Present'	'OK'	'OK'	'OK'	'OK'
23	C_023	'Not present'	'Present'	'OK'	'OK'	'OK'	'OK'
24	C_024	'Not present'	'Present'	'Warning'	'OK'	'OK'	'OK'
25	C_025	'Not present'	'Present'	'Warning'	'OK'	'OK'	'OK'
26	C_026	'Not present'	'Present'	'Warning'	'OK'	'OK'	'OK'

Table 5.
WDM investigator showing information on the fibre link characteristics.

Ch. #	Name	λ (nm)	Signal power (dBm)	OSNR (dB)	Noise (dBm)	BW 3.00 dB (nm)	BW 20.00 dB (nm)
1	C_001	1529.543	(i) 18.17	23.07	(InB)-41.24	0.232	—
2	C_002	1531.883	(i)-19.59	17.63	(InB nf)-37.22	0.138	—
3	C_003	1532.672	(i)-18.06	17.49	(InB nf)-35.55	0.132	0.391
4	C_004	1533.458	(i)-15.83	24.98	(InB)-40.81	0.130	0.299
5	C_005	1534.238	(i)-17.45	17.92	(InB nf)-35.37	0.134	0.384
6	C_006	1535.815	(i)-18.79	18.85	(InB nf)-37.64	0.068	0.313

Ch. #	Name	λ (nm)	Signal power (dBm)	OSNR (dB)	Noise (dBm)	BW 3.00 dB (nm)	BW 20.00 dB (nm)
7	C_007	1536.600	(i)-20.90	16.86	(InB nf)-37.77	0.133	—
8	C_008	1537.391	(i)-16.76	24.69	(InB)-41.45	0.128	0.302
9	C_009	1538.179	(i)-15.82	25.63	(InB)-41.45	0.125	0.276
10	C_010	1538.966	(i)-19.57	18.27	(InB nf)-37.83	0.131	—
11	C_011	1539.757	(i)-19.47	16.62	(InB nf)-36.08	0.139	—
12	C_012	1540.548	(i)-15.65	25.46	(InB)-41.11	0.131	0.296
13	C_013	1541.340	(i)-19.68	16.35	(InB nf)-36.03	0.134	—
14	C_014	1542.927	(i)-18.68	17.67	(InB nf)-36.35	0.136	0.393
15	C_015	1543.720	(i)-18.84	19.18	(InB nf)-38.02	0.135	—
16	C_016	1544.518	(i)-18.64	17.14	(InB nf)-35.79	0.131	0.378
17	C_017	1546.114	(i)-21.48	16.35	(InB nf)-37.83	0.134	—
18	C_018	1547.700	(i)-17.26	20.11	(InB nf)-37.36	0.069	0.285
19	C_019	1549.301	(i)-18.65	17.38	(InB nf)-36.03	0.065	0.323
20	C_020	1550.109	(i)-21.66	15.63	(InB nf)-37.28	0.070	—
21	C_021	1551.704	(i)-15.57	26.07	(InB)-41.64	0.068	0.258
22	C_022	1552.515	(i)-21.38	16.59	(InB nf)-37.97	0.136	—
23	C_023	1553.319	(i)-19.71	18.48	(InB nf)-38.18	0.069	0.333
24	C_024	1554.125	(i)-15.34	21.82	(InB nf)-37.16	0.131	0.285
25	C_025	1555.742	(i)-20.97	16.24	(InB)-37.21	0.068	0.373
26	C_026	1556.543	(i)-19.32	17.52	(InB nf)-36.84	0.068	0.326

Table 6.
Link spectral characterization.

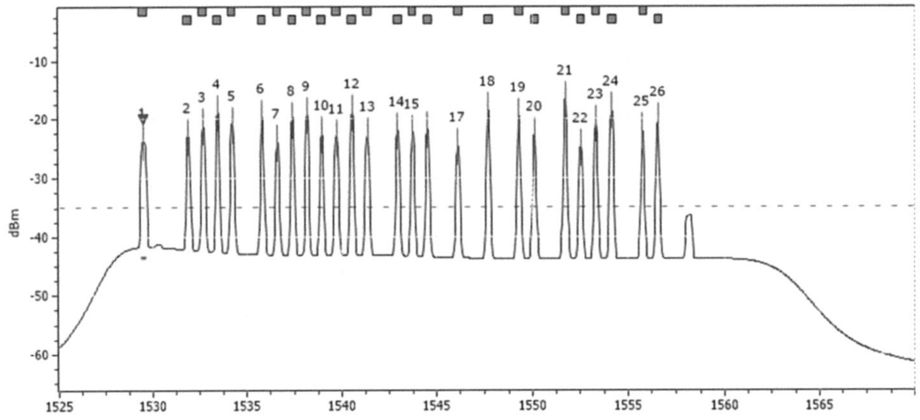

Figure 4.
WDM system characteristic of the link.

(a) Input channel								
Peak	**Channel**	**W (nm)**	**P auto (dBm)**	**SNR avg. (dB)**	**Noise avg. (dBm)**	**Pp (dBm)**	**Pi (dBm)**	**Channel Pi (dBm)**
1	100G 12	! 1537.347	5.46	45.80	...	5.46	5.26	−2.04
2	100G 24	1546.884	6.84	46.54	...	6.84	7.20	5.63
3	100G 31	1552.512	6.56	44.05	...	6.56	6.56	6.25
4	100G 39	1558.948	8.35	47.60	...	8.35	8.39	6.37
Peak	Channel	BW at 3.00 (nm)	W delta (nm)	Pp–Pavg. (dB)	Pp–Pmax (dB)	<< SNR (dB)	SNR >> (dB)	Worst SNR (dB)
1	100G 12	0.032	...	−1.20	−2.89	45.58	46.04	45.58 <<
2	100G 24	0.036	9.537	0.18	−1.51	46.64	46.45	46.45 >>
3	100G 31	0.044	5.628	−1.42	−2.93	47.09	44.20	43.91 <<
4	100G 39	0.033	6.436	1.69	0.00	50.40	47.75	47.46 <<
Average wavelength		1550.452 nm						
Average power		6.66 dBm						
Total power		12.48 dBm						
Power flatness		3.11 dB						
(b) Output channel								
Peak	**Channel**	**W (nm)**	**P auto (dBm)**	**SNR avg. (dB)**	**Noise avg. (dBm)**	**Pp (dBm)**	**Pi (dBm)**	**Channel Pi (dBm)**
1	100G 12	! 1537.347	−6.18	50.93	...	−6.18	−5.81	−12.36
2	100G 24	1546.884	−6.14	51.14	...	−6.14	−6.49	−7.80
3	100G 31	1552.509	−6.59 i	47.26	...	−7.98	−6.59	−7.00
4	100G 39	1558.948	−5.05	50.48	...	−5.05	−4.98	−7.02
Peak	Channel	BW at 3.00 (nm)	W delta (nm)	Pp–Pavg. (dB)	Pp–Pmax (dB)	<< SNR (dB)	SNR >> (dB)	Worst SNR (dB)
1	100G 12	0.036	...	0.03	−1.13	50.97	50.89	50.89 >>
2	100G 24	0.030	9.537	0.07	−1.09	51.30	50.98	50.98 >>
3	100G 31	0.045	5.625	−1.77	−3.11	43.91	47.47	437.09 <<
4	100G 39	0.034	6.439	1.16	0.00	47.46	50.56	450.56 <<
Average wavelength					1549.367 nm			
Average power					−6.21 dBm			
Total power					−0.19 dBm			
Power flatness					2.93 dB			

Table 7.
Optical spectrum analysis of EDFA at 100 GHz over 1500–1600 nm for input and output channel.

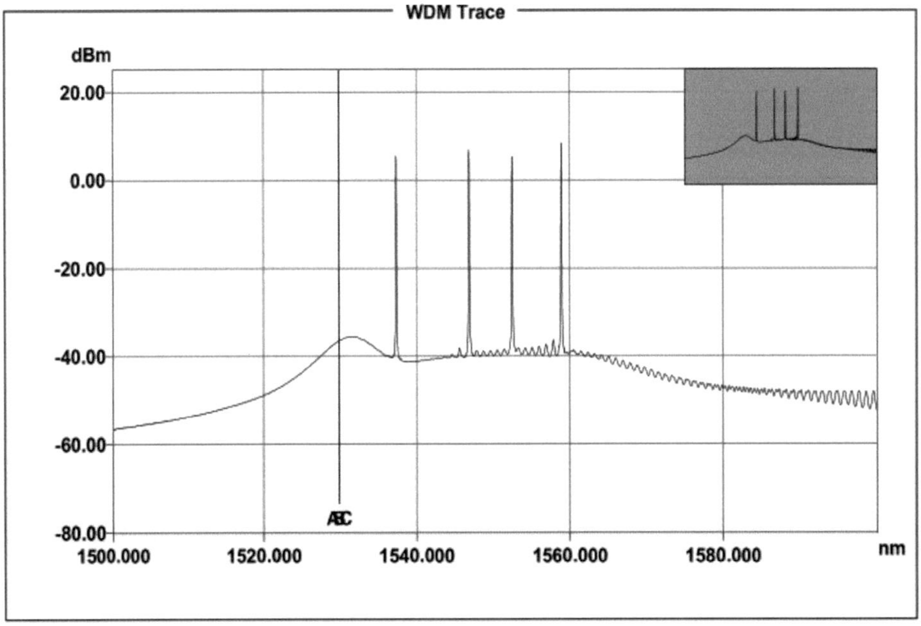

(a)

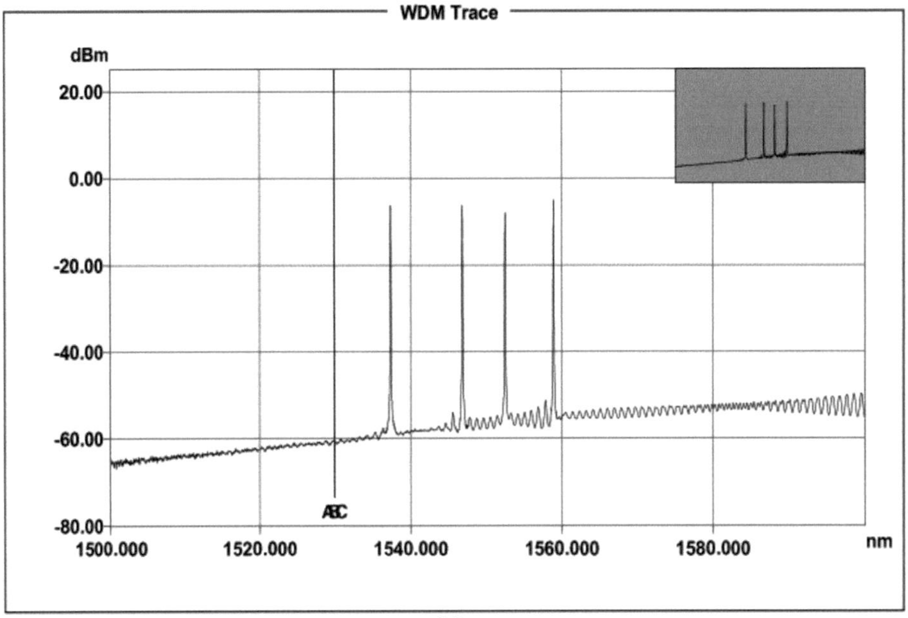

(b)

Figure 5.
OSA generated report for the EDFA at 100 GHz and 1500–1600 nm wavelength for input and output channel. (a) Input Channel, (b) Output Channel.

5. Conclusion

In this paper, it was possible to characterize long-distance fibre transmission network systems based on their spectral transmission characteristics. Fibre characterization consists of the measurement and recording of a single span parameter or multiple parameters. The characterization provides documentation for fibre parameters at the time of installation or acquisition for comparison with future measurements to determine fibre degradation due to ageing, damage and repair. This process depends on the transmission system, design margins and the reason for the measurements. At a minimum, the overall fibre loss measurements for operating wavelengths were necessary.

The analysis of the type of single-mode fibre cable for a particular transmission has been seen to be affected by the transmission wavelength. The attenuation and chromatic dispersion were similarly affected by the increase in the distance of communication.

Chromatic dispersion requires more attention in WDM systems using G.652 fibres since the dispersion was seen to be larger in the 1550 nm region. On the other hand, PMD was very noticeable at high bit rates and became alarming at bit rates in excess of 5 Gbps. PMS has been found to be the root cause of impairment in longer-distance optical WDM systems. Hence, to avoid this kind of impairment for transmission systems operating at a bitrate higher than 10 Gbps, PMD fibre compensators with a certain degree of birefringence must be employed to further reduce the impairment to the barest minimum.

An optical spectrum analyzer was used to measure the distribution of optical power energy across the wavelength channel enabling spectral analyses, monitoring of optical signals, assessment of optical amplifier, network analysis and OSNR measurement.

Author details

Asiya E. Asiya[1*], Michael U. Onuu[2], Rufus C. Okoro[3] and O. Enendu Uche[4]

1 Electronics and Computer Technology Unit, University of Calabar, Calabar, Cross River State, Nigeria

2 Engineering Physics Research Group, Department of Physics/Geology/Geophysics, Alex Ekwueme Federal University, Ndufu-Alike, Ebonyi State, Nigeria

3 Faculty of Physical Sciences, Department of Physics, University of Calabar, Calabar, Nigeria

4 Intelligence Information Processing Technology, Department of Computer Engineering, Huzhou University, China

*Address all correspondence to: evangephraim@yahoo.co.uk

References

[1] Govind PA. Fibre-Optic Communication Systems. Rochester, NY: John Wiley; 2021

[2] Zhao T, Wang A, Wang Y, Zhang M, Chang X, Xiong L, et al. Fibre fault location utilizing traffic signal in optical network. Optics Express. 2013;**21**(20): 23978-23984

[3] Song H, Kim BW, Mukherjee B. Long-reach optical access networks: A survey of research challenges, demonstrations, and bandwidth assignment mechanisms. IEEE Communications Surveys & Tutorials. 2010;**12**(1):112-123

[4] Stremler FG. Introduction to Communication Systems. Boston, USA. 1990

[5] Eriksson TA, Bülow H, Leven A. Applying neural networks in optical communication systems: Possible pitfalls. IEEE Photonics Technology Letters. 2017;**29**(23):2091-2094

[6] Kumar V, Rajouria D. Fault detection technique by using OTDR: Limitations and drawbacks on practical approach of measurement. IJETAE. 2012;**2**:283-287

[7] Kyriakopoulos CA, Papadimitriou GI, Nicopolitidis P, Varvarigos E. Energy-efficient lightpath establishment in backbone optical networks based on ant colony optimization. Journal of Lightwave Technology. 2016;**34**(23): 5534-5541

[8] Tanimura T, Hoshida T, Rasmussen JC, Suzuki M, Morikawa H. OSNR monitoring by deep neural networks trained with asynchronously sampled data. In: 2016 21st OptoElectronics and Communications Conference (OECC) Held Jointly with 2016 International Conference on Photonics in Switching (PS). Japan: IEEE; 2016. pp. 1-3

[9] Zhang X, Lu F, Chen S, Zhao X, Zhu M, Sun X. Remote coding scheme based on waveguide Bragg grating in PLC splitter chip for PON monitoring. Optics Express. 2016;**24**(5):4351-4364

[10] Hayford-Acquah T, Asante B. Causes of fiber cut and the recommendation to solve the problem. IOSR Journal of Electronics and Communication Engineering. 2017;**12**(1):46-64

[11] Pournoury M, Moon DS, Nazari T, Kassani SH, Do MH, Lee YS, et al. Low scattering loss fiber with segmented-core and depressed inner cladding structure. Optics Communications. 2014;**317**:13-17

[12] El-Sayed M, Ibrahim PJ, Gunzer F. Investigation of the precision regarding fiber fault location with a commercial optical time domain reflectometer. In: 7th International Symposium on High-Capacity Optical Networks and Enabling Technologies. United States: IEEE; 2010. pp. 237-241

[13] Cohen E, Malka D, Shemer A, Shahmoon A, Zalevsky Z, London M. Neural networks within multi-core optic fibers. Scientific Reports. 2016;**6**(1):1-14

[14] Mo W, Huang YK, Zhang S, Ip E, Kilper DC, Aono Y, et al. ANN-based transfer learning for QoT prediction in real-time mixed line-rate systems. In: 2018 Optical Fiber Communications Conference and Exposition (OFC). California, USA: IEEE; 2018. pp. 1-3

[15] Shiryaev VS, Karaksina EV, Churbanov MF, Kotereva TV, Stepanov BS, Ketkova LA, et al. Special pure germanium-rich Ga-Ge-As-Se glasses

for active mid-IR fiber optics. Materials Research Bulletin. 2018;**107**:430-437

[16] Soujanya A, Goud OSC, Prasad KS, Reddy GP. Featured based pattern analysis using machine learning and artificial intelligence techniques for multiple featured dataset. Materials Today: Proceedings. 2017;**4**(8): 9039-9048

[17] Rubio-Largo A, Vega-Rodriguez MA, GomezPulido JA, Sanchez-Perez JM. A comparative study on multi-objective swarm intelligence for the routing and wavelength assignment problem. IEEE Transactions on Systems, Man and Cybernetics, Part C (Applications and Reviews). 2012;**42**(6):1644-1655

[18] Shahkarami S, Musumeci F, Cugini F, Tornatore M. Machine-learning-based soft-failure detection and identification in optical networks. In: 2018 Optical Fiber Communications Conference and Exposition (OFC). IEEE; 2018. pp. 1-3

[19] Behrens C. Mitigation of Nonlinear Impairments for Advanced Optical Modulation Formats. United Kingdom: UCL (University College London); 2012

[20] Xie C, Möller L. The accuracy assessment of different polarization mode dispersion models. Optical Fiber Technology. 2006;**12**(2):101-109

[21] Jim H. The Reference Guide to Fibre Optics. Santa Monica, CA: The FOA; 2011

[22] Xie F. Optical fibre Line Failure Detecting. Fínland: Vaasa University of Applied Sciences VAMK; 2013

[23] Luo ZQ, Zhang S. Dynamic spectrum management: Complexity and duality. IEEE Journal of Selected Topics in Signal Processing. 2008;**2**(1):57-73

[24] Thongdaeng R, Worasucheep DR. Effect of bending radius and bending location on insertion loss in single mode fibers and polarization maintaining fibers. Procedia Computer Science. 2016; **86**:15-18

[25] Wass J, Thrane J, Piels M, Jones R, Zibar D. Gaussian process regression for WDM system performance prediction. In: 2017 Optical Fiber Communications Conference and Exhibition (OFC). Los Angeles, California, USA: IEEE; 2017. pp. 1-3

[26] Keiser G. Fiber Optic Communications. Massachusetts Ave., NW: Springer Nature; 2021

[27] Urban PJ, Getaneh A, Von der Weid JP, Temporão GP, Vall-llosera G, Chen J. Detection of fiber faults in passive optical networks. Journal of Optical Communications and Networking. 2013;**5**(11):1111-1121

[28] Cen M, Chen J, Mégret P, Moeyaert V, Wuilpart M. Fast and simple fault monitoring for long-reach passive optical networks. In: 2014 The European Conference on Optical Communication (ECOC). Cannes, France: IEEE; 2014. pp. 1-3

[29] Cen M, Chen J, Moeyaert V, Mégret P, Wuilpart M. Multi-wavelength transmission-reflection analysis for fiber monitoring. Optics Express. 2014;**22**(25): 31248-31262

[30] Shtaif M, Mecozzi A, Nagel JA. Mean-square magnitude of all orders of polarization mode dispersion and the relation with the bandwidth of the principal states. IEEE Photonics Technology Letters. 2000; **12**(1):53-55

[31] Park J, Baik J, Lee C. Fault-detection technique in a WDM-PON. Optics Express. 2007;**15**(4):1461-1466

[32] Russell E. Optical Reflections in High-Capacity DWDM Systems, CorningWhite Paper, New York, USA: WP7140. 2020

[33] Collings B, Heismann F, Lietaert G. Reference Guide to Fiber Optic Testing. California, USA: JDS Uniphase Corporation; 2010

[34] https://www.explainthatstuff.com/fiberoptics.html

[35] Makokha J, Odhiambo J. Optical characterization of atmospheric aerosols via airborne spectral imaging and self-organizing map for climate change diagnostics. Open Access Library Journal. 2018;5:1-10. DOI: 10.4236/oalib.1104698

Chapter 3

A Comparative Analysis of Linear Modulation Schemes in the Presence of FLI in Wireless Optical Systems

Aruna Tripathy and Aishwarya Dash

Abstract

In this chapter three optical modulation techniques such as on off keying (OOK), binary phase shift keying (BPSK), differential phase shift keying (DPSK) are evaluated in terms of Q-factor and bit error rate (BER) in order to assess their suitability in typical indoor optical wireless communication. The modulated signals are subjected to additive white Gaussian noise (AWGN) and florescent light interference (FLI) representative of practical scenarios. The received signals are demodulated with and without using matched filter (MF). The BER in both the cases are evaluated through simulation studies using MATLAB and compared for each of the three modulated signals. It has been observed that in each of the demodulation process the rate of reduction of BER is significant when the demodulation is done by using MF.

Keywords: OOK, BPSK, DPSK, FLI, AWGN, matched filter, BER, SNR

1. Introduction

The day to day increasing number of wireless devices has drastically increased the crowd in radio frequency channels. Starting from 1G to 4G the objective of development has been to increase the data rate and channel reusability. This is due to the fact that the Radio Frequency (RF) band assigned for communication is quite limited while the applications and user number are increasing day by day. The inter-channel interference and electromagnetic interference (EMI) are two impairments of the RF channel. Hence an alternative needs to be investigated in order to meet the demand of the growing number of users. Question arises as to the suitability of channel that can be a replacement of such a highly reliable and high data rate supporting channel (?). Optical communication is a promising technique to provide a reliable and high speed data transmission. Again fiber optic cables are free of EMI. The major issue associated with optical fiber cable is pulse dispersion which limits the data rate of the signal to be transmitted through it. Also this dispersion increases with increasing distance. In rural areas, areas with rough, difficult to access terrains, the installation and maintenance of optical fiber cables is not commercially viable. For this reason wireless radio channels are becoming important as an alternative. With increasing applications of wireless devices, the demand for radio channels is also increasing, but the channel band width is limited. Different frequency reuse techniques are being developed as a potential solution. Another solution has been to use optical wireless channel as an alternate option of radio channel, so that the crowd in radio frequency can be reduced. It may be considered to be a hybrid of optical and wireless technology.

Wireless optical communication is a technique to transmit and receive optically modulated signals in free space or air [1]. So the frequency band that the signal uses is the optical band. But ambient light interference is a major problem while optical signal is transmitted through wireless medium [2]. This necessitates solutions to have to be found out to make optical wireless communication a practical potential candidate.

Optical communication techniques use different modulation techniques where some techniques are power efficient and some others are spectrally efficient. The spectral efficiency of a modulated signal depends upon the modulation scheme as well as the line-coding technique. In [3] a comparative study has been carried out between the performance of On-Off Keying (OOK), Pulse Position Modulation (PPM) and Digital Pulse Interval Modulation (DPIM) systems in presence of florescent light interference (FLI) by considering normalized optical power required (NOPR) and optical power penalty (OPP) as the performance metrics. A 64 channel inter-satellite wireless optical channel (IS-OWC) based system has been described in [4], in which three most preferable modulation techniques such as Alternate Mark Inversion (AMI), Differential Phase Shift Keying (DPSK) and Chirp-RZ have been compared on the basis of their Q-factors, BER and eye diagram as achieved at different data rates. Their spectra were compared based on their bandwidth. The results reveal DPSK as the best among all. By using space and polarization diversity techniques in [5], a high speed IS-OWC system has been designed and simulated. The analysis has been made on the metrics like BER, Q-factor, eye opening etc. The aim is to find out the practicability of OWC for establishing an inter satellite communication. In [6], advanced modulations, multiplexing, coding schemes, detection schemes, switching and controlling schemes as applicable in typical optical communication have been discussed with possible directions for making further research. In [7] an experimental analysis has been done for OOK-return to zero (RZ) and non return to zero (NRZ) signal in a Light Fidelity (Li fi) network and it has been found that OOK-NRZ has a wider coverage area as compared to that of OOK-RZ. In [8] performance of RZ modulation has been found to be better than NRZ modulation in Gigabit Passive Optical Network (GPON) system. The OOK-RZ and NRZ signals have been analyzed over Free Space Optical (FSO) channel in different weather conditions in [9] wherein the analysis is carried out by taking different wireless media. [10–13] gives analytical observations on some advanced modulation techniques such as FSK, DPSK, QDPSK based on RZ and NRZ pulses.

In this chapter OOK, BPSK and DPSK optical transreceiver systems are designed with an objective of supporting high data rate (as needed for 5G communication), low energy consumption, high received power and an affordable design complexity. The simulation of the above three systems are first done by using Optisys16.0 by taking both RZ and NRZ line coding for a certain set of parameters to find which type of line coding will give a better result. Then the performance of each system has been investigated in presence of FLI and AWGN while considering the BER as the performance parameter. Simulation studies are presented to show the superior BER achieved through the use of a matched filter (MF) as compared to no MF at the receiver.

DOI: http://dx.doi.org/10.5772/intechopen.109841

2. System components

Here the intention is to transmit the signal in free space. Again the signal which will be received at the other end that will be an optical signal and hence the antenna must be an optical antenna. The receiver must be capable to convert the optical signal into the detectable form. So there are some special system components used to perform all these operations.

2.1 Wireless optical channels

In any communication process, channel plays an important role. The channel performance characteristics somehow depend on the design of the transmitting and receiving antenna. Wireless optical channel includes the frequency range of infra-red, visible light and ultraviolet [1]. The optically modulated signal is radiated to freespace or air at the transmitter side. At the receiver side, the optical signal is accumulated by using telescope. In this analysis Optical Wireless Channel (OWC) has been used, however another technique named Free Space Optics (FSO) is also being used in some fields. The equation of wireless optical channel is same as RF channel [4].

Wireless optical links are of two types—Optical Wireless Channel (OWC) and Free Space Optical Channel (FSO) [1]. The differentiation has been established between the two channels by considering the difference between the transmitting and receiving antenna used for the two techniques. OWC can be said to be as an improved version of FSO [5]. FSO needs LOS path and it works effectively for short range [1, 14]. For long distance communication OWC is used.

Figure 1 represents an FSO based system where the transmitting antenna is simply a light radiator and the receiving antenna is a photo detector which has the ability to convert the photonic power into voltage signal. The idea of FSO is based on the

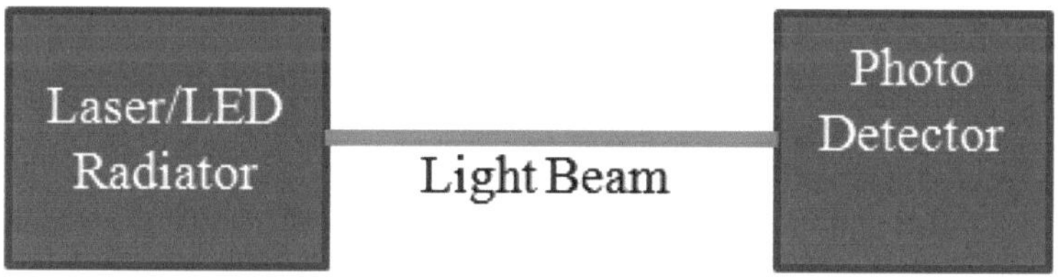

Figure 1.
Block schematic of a simple FSO based system.

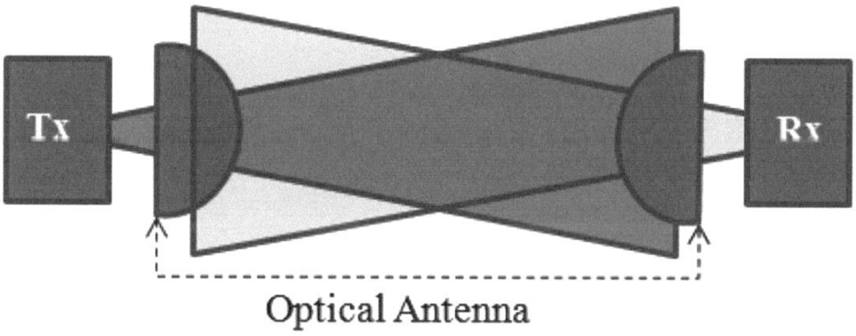

Figure 2.
Block schematic of a simple OWC based System.

transmission of collimated light beam from one location to another by using low power IR laser [1].

Figure 2 represents a simple block diagram of OWC system. The concept of OWC involves the use of single mode fibers directly as light launchers and light collectors.

2.2 Continuous wave (CW) laser

CW laser is being used as the optical source. Continuous wave means that it is continuously pumped and emits light continuously, not in form of pulses [1].

2.3 Mach-Zehnder (MZ)-modulator

A Mach-Zehnder modulator is simply an interferometer [15, 16]. It is otherwise known as Mach-Zehnder Interferometer (MZIM). It is a combination of two electro-optic phase modulators (EOPM) in parallel structure as shown in **Figure 3**. It splits the input optical signal into two fields by the two arms of the MZIM and at the output end these two fields are again recombined by means of interference.

MZIM is again of two types—single drive and dual drive. In single drive MZIM one arm is driven by an input voltage $\boldsymbol{V(t)}$ while no voltage is applied to the other arm. Therefore the output optical field is given by [16]

$$E_0(t) = \frac{E_i(t)}{2}\left[1 + e^{j\pi\frac{(V(t)+V_{bias})}{V_\pi}}\right] = F_i(t)\cos\left[\frac{\pi}{2}\frac{(V(t)+V_{bias})}{V_\pi}\right] e^{-j\left[\frac{\pi}{2}\frac{(V(t)+V_{bias})}{V_\pi}\right]} \quad (1)$$

where

$\boldsymbol{E_i(t)}$ = Input optical field.

$\boldsymbol{V_{bias}}$ = DC voltage.

$\boldsymbol{V_\pi}$ = Driving voltage required to create a $\boldsymbol{\pi}$ phase shift on the light wave carrier (usually 3–6 V).

The dual drive MZIM takes two driving voltages, $\boldsymbol{V_1(t)}$ and $\boldsymbol{V_2(t)}$ such that $\boldsymbol{V_2(t) = -V_1(t)}$. Therefore, considering $\boldsymbol{V_1(t) = V(t)}$ and $\boldsymbol{V_2(t) = -V(t)}$, the output optical field is given by [16]

$$\boldsymbol{E_0(t) = \frac{E_i(t)}{2}\left[e^{j\pi\frac{(V(t)+V_{bias})}{V_\pi}} + e^{j\pi\frac{-(V(t)+V_{bias})}{V_\pi}}\right]}$$

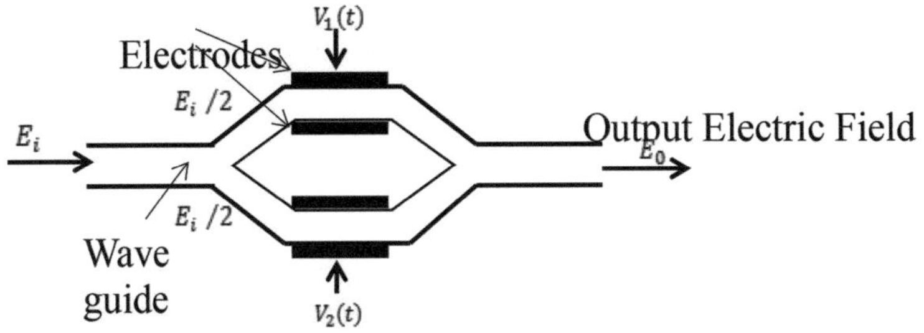

Figure 3.
MZ interferometer.

$$= E_i(t) \cos \left[\frac{\pi}{2} \frac{(V(t) + V_{bias})}{V_\pi} \right] \quad (2)$$

In all these analysis Dual-drive MZ-modulator has been used however single-drive MZIM can also be used with the same arrangement.

2.4 Avalanche photodiode (APD)

APD is a highly sensitive photodiode [15, 16] that is used to convert the photonic energy into electrical energy. As compared to PIN diode it can detect lower level light energy. Hence these diodes are basically used at the receiver side in long distance transmission. APD provides inherent current gain through repeated electron ionization process. Hence a multiplication factor **M** is associated with the expression of responsivity of an APD which is given by

$$M = I_T / I_P \quad (3)$$

where, $\mathbf{I_T}$ = total output current; $\mathbf{I_P}$ = primary photocurrent.

3. Optical OOK, BPSK and DPSK modulation systems

The signal wave forms and the logical statements of the above said optical systems are quite similar to that of the electrical systems but the functional blocks of the optical systems are completely different from the electrical systems. It is obvious that the system components used here must support the optical signals. So here the system block diagrams are explained well with their corresponding Optisys16.0 simulator set up that has been used here to analyze the performance of the systems.

3.1 Optical ON-OFF keying (OOK)

It is the simplest modulation technique used in optical modulation schemes. The binary data is first converted into NRZ pulse then given to the MZIM as the driving voltage. The optical signal from CW laser is given to the input of the modulator. In the output the ON-OFF keying optical signal is generated. **Figure 4** shows the block schematic of a typical OOK modulator. **Figure 5** represents a simple optical receiver in which a single Avalanche photo diode (APD) has been used to convert the optical

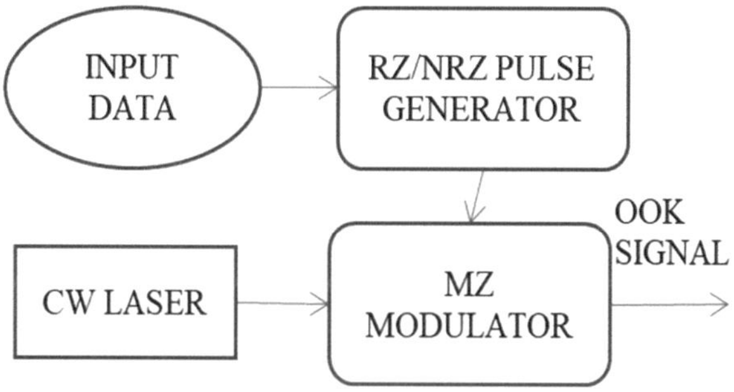

Figure 4.
ON-OFF keying modulator.

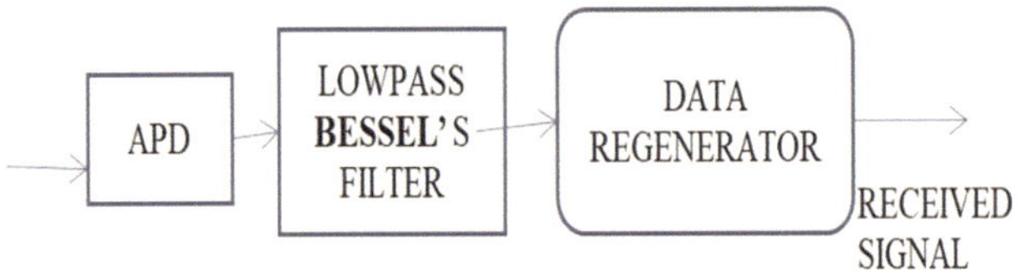

Figure 5.
ON-OFF keying demodulator.

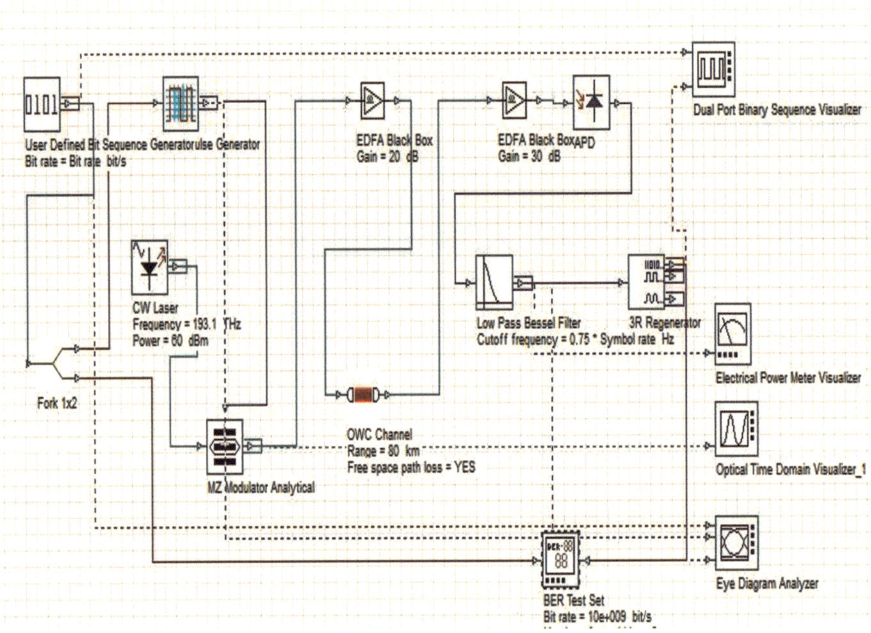

Figure 6.
Simulation layout of OOK Transreceiver using OWC in Optisys16.0.

signal into electrical signal. Then the signal is passed through a LPF to get the desired frequency. The data regenerator will generate the binary signal from the electrical signal. **Figure 6** shows the entire setup in Optisys16.0 to simulate the OOK system with OWC channel.

3.2 Optical binary phase shift keying

Here the binary data is first sent to BPSK pulse generator and then given to the MZIM as the driving voltage. The optical signal from CW laser is given to the input of the modulator. In the output the BPSK optical signal is generated. **Figure 7** shows the block schematic of a typical BPSK modulator. **Figure 8** represents a simple optical receiver in which a single PIN diode has been used to convert the optical signal into electrical signal. Then the signal is passed through a LPF to get the desired frequency. The data regenerator will generate the binary signal from the electrical signal. **Figure 9** shows the entire setup in Optisys16.0 to simulate the BPSK system with OWC channel.

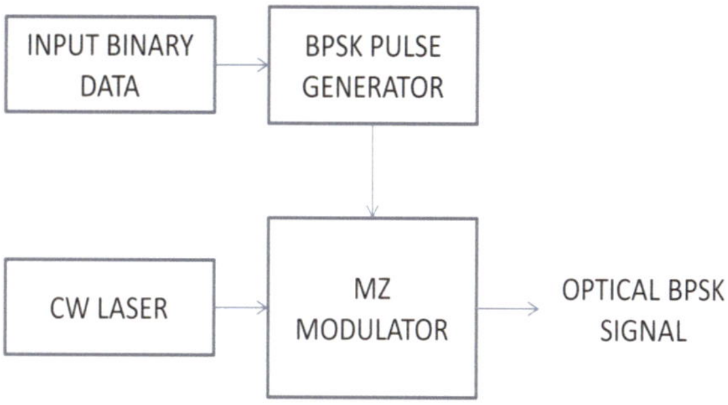

Figure 7.
BPSK modulator.

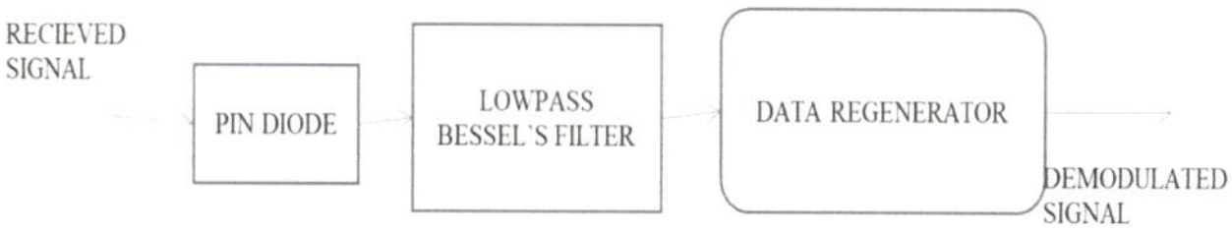

Figure 8.
BPSK demodulator.

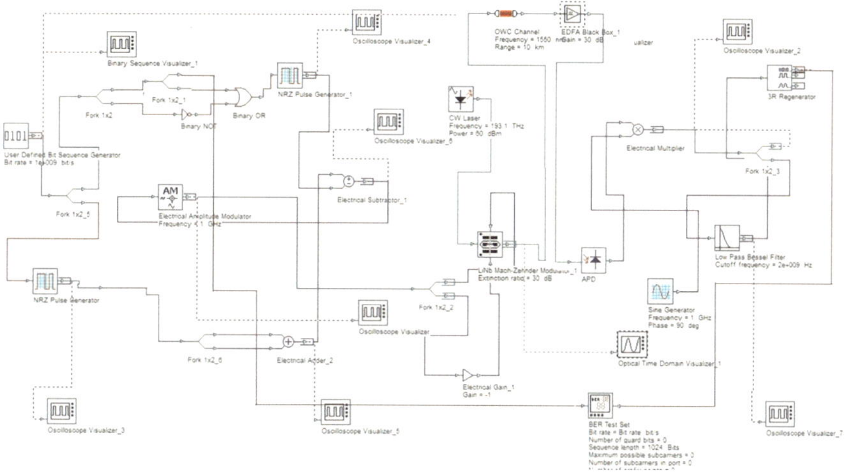

Figure 9.
Simulation layout of BPSK Transreceiver using OWC in Optisys16.0.

3.3 Optical differential phase shift keying (ODPSK)

Figure 10 represents an DPSK modulator. The binary signal is first converted into a Duobinary code by using a duo binary pre-coder. The internal circuitry of a Duobinary precoder has been given in **Figure 11** which is simply a two input XOR gate having one input connected to the output and the other input is the binary data to be encoded. The initial value of D is set to 1, then from the next step the output is fed back to the input. An example of te conversion is done in **Table 1**. The signal conversion from binary to duobinary has been shown in **Figures 12** and **13**. Thus, the duo-binary coded data is generated which is then fed to the pulse generator. Then the pulse is given to the first modulator which converts the electrical signal into optical

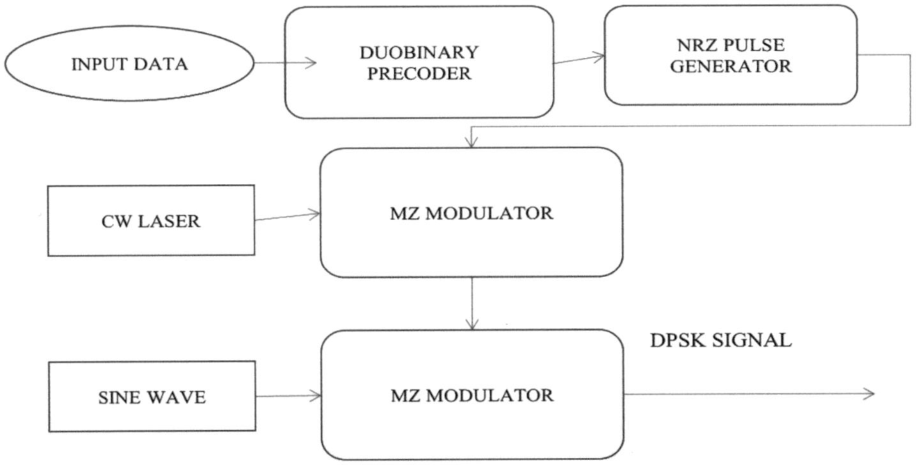

Figure 10.
DPSK modulator.

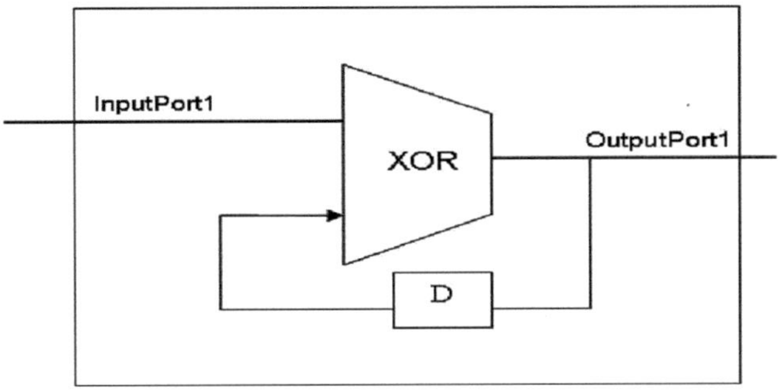

Figure 11.
Duobinary Precoder.

d_k	D	b_k
1	1	0
0	0	0
0	0	0
1	0	1
1	1	0
0	0	0
1	0	1
1	1	0
1	0	1
0	1	1

Table 1.
Duobinary coding.

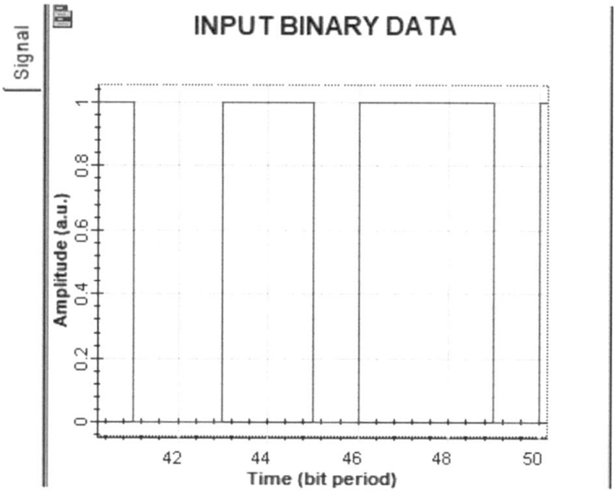

Figure 12.
Input signal to Duobinary Precoder.

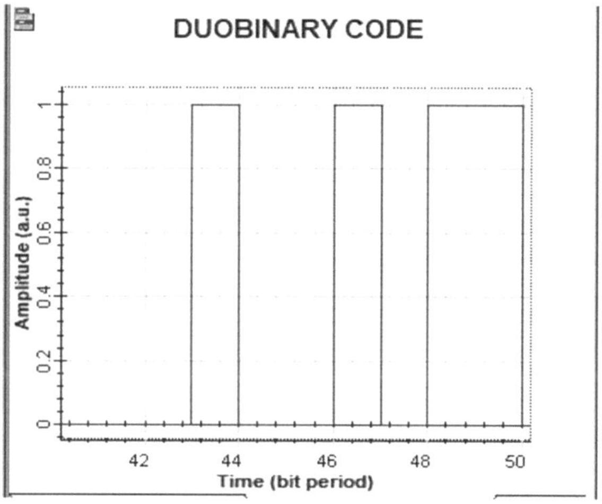

Figure 13.
Output of Dubinary Precoder.

signal. The second modulator take the driving voltage from a sinusoidal function generator that finally gives a differerentially phase shifted optical signal or optical DPSK signal.

Figure 14 represents the block schematic of optical DPSK demodulator where the optical receiver block consists of an APD and a DPSK decoder. The signal is then passed through an LPF and then the binary data is generated from the electrical signal by the data regenerator.

Figure 14 represents the block schematic of optical DPSK demodulator where the optical receiver block consists of an APD and a DPSK decoder. The signal is then passed through an LPF and then the binary data is generated from the electrical signal by the data regenerator (**Figure 15**).

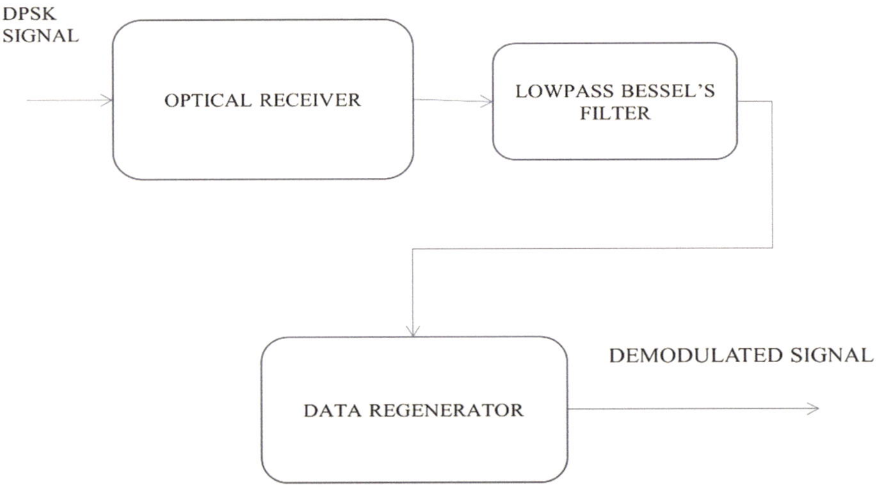

Figure 14.
DPSK-demodulator.

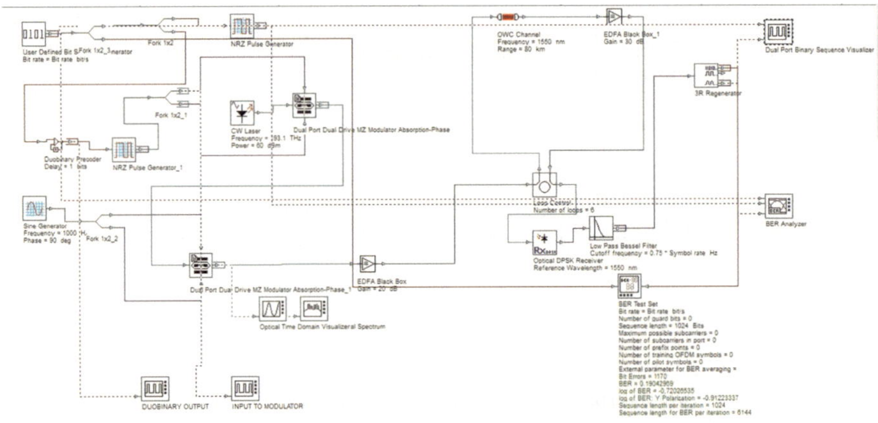

Figure 15.
Simulation layout of optical DPSK transreceiver using OWC in Optisys16.0.

4. Effect of ambient light interference

Sun light causes the maximum amount of background noise and it is the major source of shot noise at the receiver. The ambient light has an operating wavelength range of 850 nm to 1550 nm. It can be modeled as a white Gaussian noise with PSD given by [1].

$$N_0 = 2qI_b \tag{4}$$

where q = charge of electron; I_b = amount of current.

Incandescent lamps have a maximum Power Spectral Density (PSD) around 1 μm. It can produce an interference of near perfect sinusoid with a frequency of 100 Hz [1]. So the first few harmonics carry a significant amount of energy which can be avoided by using an High Pass Filter (HPF).

Fluorescent light has a switching frequency range in between 20 and 40 kHZ [17, 18]. The detected electrical signal may contain harmonics of the switching frequency. These harmonics may extent to the range of Megahertz frequency range and can create a much more serious impairment to the receiver in case of wireless optical system [19, 20]. When a signal is received along with channel noise and also FLI, the signal cannot be detected only by thresholding. The received signal has an inherent delay in reaching the receiver and is also subject to amplitude distortion occurs due to FLI besides the omnipresent additive noise. All these render an unacceptable signal to noise ratio (SNR) at the receiver input. So it becomes difficult to sample the signal at an accurate instant. Hence it becomes impossible to detect the signal with acceptable accuracy. In this case a special kind of matched filter is to be used. The formulation of this filter is described in the next section.

5. Matched filter

The matched filter is an optimal linear filter for maximizing the SNR in presence of additive noise. It is used to detect highly noisy signal [21].

Let us consider the general equation of a received signal,

$$R_k = T_k + n_k \tag{5}$$

where

$\boldsymbol{R_k}$ = received signal.

$\boldsymbol{n_k}$ = additive noise.

$\boldsymbol{T_k}$ = scaled and shifted version of a known signal

$$T_k = \mu_0 S_{k-j_0} \tag{6}$$

$\boldsymbol{\mu_0}$ = unknown scaling.

$\mathbf{j_0}$ = unknown shift.

To detect the original signal the optimal estimation of $\boldsymbol{j_0}$ and $\boldsymbol{\mu_0}$ is needed.

Let the optimal estimation of $\boldsymbol{j_0}$ and $\boldsymbol{\mu_0}$ be $\boldsymbol{j}$ and $\boldsymbol{\mu}$ respectively which are obtained by minimizing the least square residual between the observed sequence $\boldsymbol{R_k}$ and probing sequence $\boldsymbol{h_{j-k}}$. This $\boldsymbol{h_{j-k}}$ is otherwise the impulse response of the matched filter.

$$j^*, \mu^* = \arg\min_{j,\mu} \sum\nolimits_k \left(R_k - \mu.h_{j-k}\right)^2 \tag{7}$$

By solving the equations the values of $\boldsymbol{\mu}$, $\boldsymbol{j}$ and $\boldsymbol{h_{j-k}}$ are found to be

$$\therefore \mu^* - \frac{\sum_k T_k h_{j-k}}{\sum_k h_{j-k}^2} \tag{8}$$

$$j^* = arg\max_j \frac{\left(\sum_k T_k h_{j-k}\right)^2}{\sum_k h_{j-k}^2} \tag{9}$$

$$h_{j-k} = \nu.T_k = \kappa.S_{k-j_0} \tag{10}$$

Hence, the probing signal h_{j-k} is proportional to S_{k-j_0}.

To detect the correct signal the observed signal must be convolved with the matched filter impulse response.

If the signal received by the receiver including FLI is given by $\boldsymbol{R_k(t)}$, then output of the matched filter is given by

$$m_0(t) = R_k(t) \otimes h_{j-k}(t) \tag{11}$$

i.e. the convolution of the received signal and the impulse response of the matched filter.

For the system here under observation $\boldsymbol{T_k}$ is the signal transmitted in the form of OOK or BPSK or DPSK where the additive noise includes both AWGN as well as FLI.

The FLI interference noise power added to the transmitted signal is given by [20]

$$n_{FLI} = R_a q \int_{-\infty}^{\infty} p(\tau) h_I^2(t-\tau) d\tau \tag{12}$$

$\boldsymbol{R_a}$ = responsivity of the receiver
$\boldsymbol{q}$ = electronic charge
$\boldsymbol{h_I(t)}$ = receiver impulse response

$$p(t) = H(t) A_r$$

$\boldsymbol{H(t)}$ = irradiance produced by the background light at the receiver side
$\boldsymbol{A_r}$ = photodetector response area
For the sampled version of transmitted signal the interference also occurs at every sample.

6. Proposed system model

Here three different types of modulation techniques are investigated on basis of their BER in presence of Gaussian noise as well as FLI. The modulation schemes chosen here are OOK, BPSK and DPSK. **Table 1** represents the simulation parameters considered during the simulation of the system model.

Figure 16 is a simple block schematic representation of the proposed system. The modulator is OOK, BPSK and DPSK. Here the noise has been added in two different

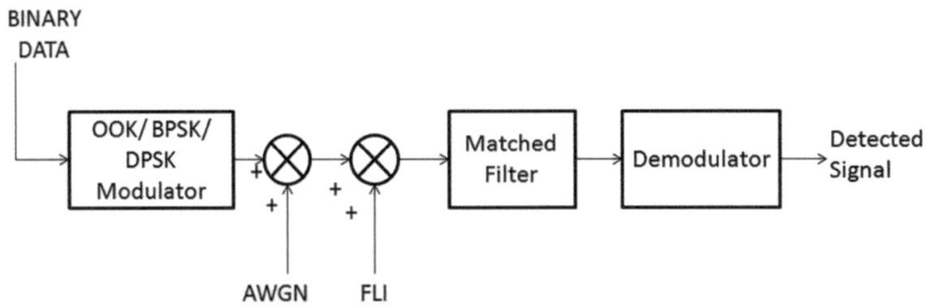

Figure 16.
Block diagram of the proposed system.

stages as shown in this figure. The received signal is passed through a matched filter before entering the demodulator.

7. Experimental outputs of the systems in Optisys16.0 simulator for Q-Factor comparison of OOK, BPSK and DPSK Transreceivers at Different Data Rates

The simulation has been done in Optisys16.0 by taking the parameters as given in **Table 2**. Then by changing the data rates from 10 Gbps to 20 Gbps and 40 Gbps the corresponding BER and Q-factor has been noted down in **Table 3**.

SL. no.	Parameters	Values
1	Channels/bits	64
2	Bit rates	10, 20, 40 Gbps
3	Input power	30 dBm
4	EDFA gain	20 dBm
5	Length of sequence	128
6	Range	2500 km
7	Samples per bit	64

Table 2.
Parameters to simulate Q-factor and BER at different data rates.

Modulation technique	Data rate (in Gbps)	Q-factor	BER
OOK	10	6.67	0.004998
	20	5.472	0.00886
	40	2.24	0.06324
BPSK	10	19.21842	0.00005321
	20	5.94129	0.00002932
	40	4.03944	0.007106
DPSK	10	24.4962	0.00006759
	20	9.1693	0.0003917
	40	5.6349	0.000712

Table 3.
Q-factor and BER at different data rates.

8. Distance versus received power for signal transmitted in RZ and NRZ linecoding

Table 4 represents observed readings of the power meter at receiver side. It shows the received signal power at different distances for all the three modulated signals

Sl no.	Distance in KM	Received power (in dBm)					
		OOK-NRZ	OOK-RZ	BPSK-NRZ	BPSK-RZ	DPSK-NRZ	DPSK-RZ
1	100	22.981	20.882	22.321	21.148	24.846	23.115
2	200	22.976	20.863	22.312	20.940	24.842	23.106
3	300	22.832	20.849	22.315	20.911	24.837	23.080
4	400	22.800	20.821	22.296	20.888	24.824	23.066
5	500	22.723	20.699	22.284	20.769	24.818	23.054
6	600	22.673	20.560	22.277	20.660	24.811	23.031
7	700	22.566	20.342	22.273	20.557	24.778	23.007
8	800	22.392	20.318	22.269	20.497	24.759	22.991
9	900	22.188	20.298	22.251	20.311	24.742	22.766
10	1000	21.889	20.008	21.211	19.967	24.728	22.611

Table 4.
Received powers for varying link length.

Sl no.	Parameters	Value
1	Data rate	10 Gbps
2	Input power	30 dBm
3	Center frequency	193.1THz
4	Data length	1024 bits
5	Link length	100 km to 1000 km

Table 5.
Parameters for simulating at varying distance.

taking both RZ and NRZ pulses. In every case the received power for NRZ signal is more than that of RZ signals. Here in every modulation schemes, the received power for NRZ modulation is more as compared to RZ modulation. The simulation is done by taking the parameters as given in **Table 5**.

9. Experimental outputs of the system with FLI

The FLI effect is analyzed by taking the system parameters as given in **Table 6**. To visualize the system performance clearly an 8-bit signal is taken and the test results for OOK based system are shown below.

Figure 17 represents the original OOK signal which has been transmitted from the transmitter. This is passed through the OWC channel to reach the receiver side. **Figure 18** represents the received signal without adding the FLI. Here the signal is noisy due to the addition of AWGN. However, as here the aim is to investigate the effect of fluorescent light interference in addition to AWGN, the FLI has also been added. The

Sl. no	Parameter	Values
1	Length of data	1024
2	Normalized data rate	1, 10 and 20 Gbps
3	SNR	1–40 dB
4	Responsivity of receiver	1

Table 6.
Simulation parameter values for BER analysis.

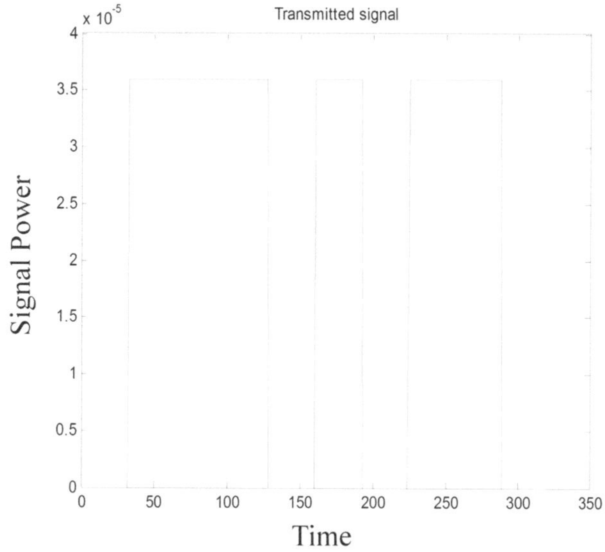

Figure 17.
Transmitted signal.

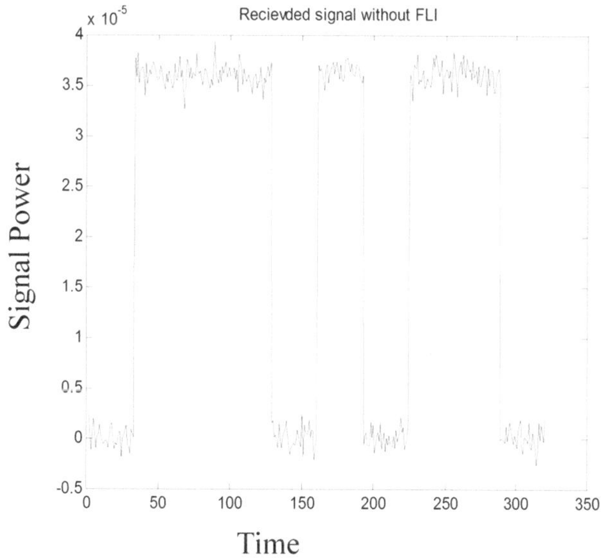

Figure 18.
Received signal without FLI.

signal becomes more distorted as shown in **Figure 19**. The noisy received signal in the presence of AWGN only can be easily detected by an integrate and dump filter; low pass filter (LPF) and subsequent thresholding. But the signal shown in **Figure 18** when down sampled, filtered and thresholded, yields the signal as shown in **Figure 20**.

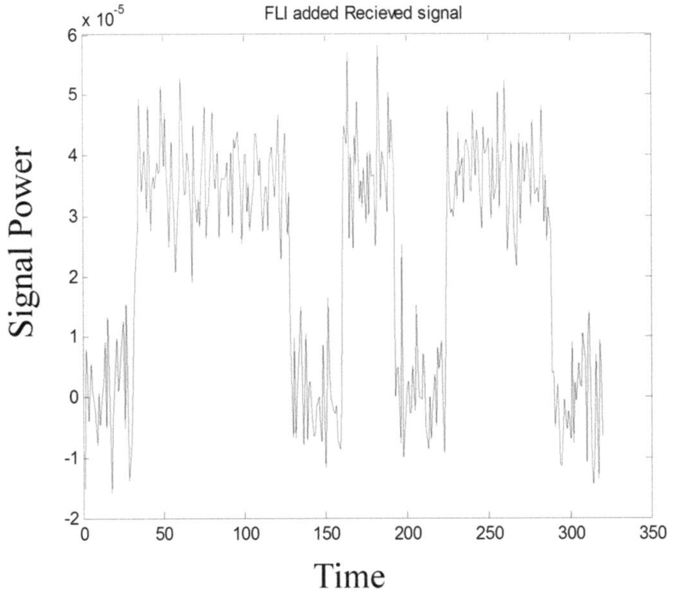

Figure 19.
Received signal with FLI.

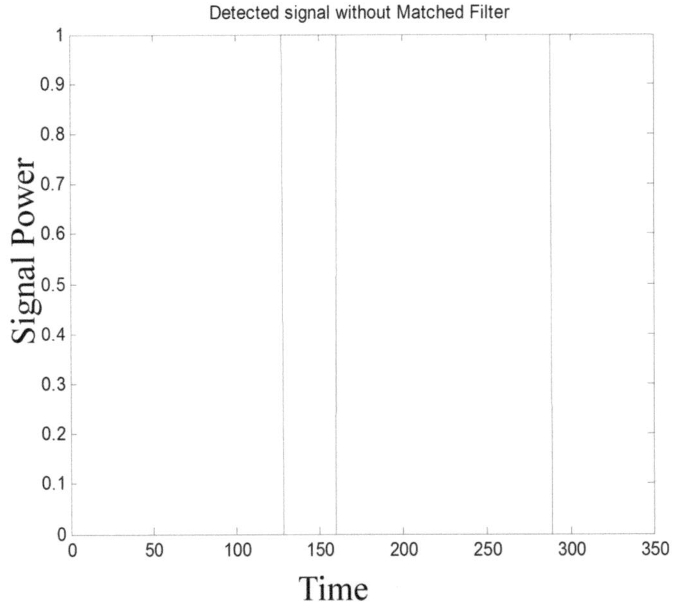

Figure 20.
Detected signal without MF.

As we can observe, the signal so recovered does not properly match with the - transmitted signal. However, when the signal is passed through the MF, where it is also down sampled, the output looks as given in **Figure 21**. By thresholding the output of MF, the signal so recovered is shown in **Figure 22**, which agrees with the transmitted signal.

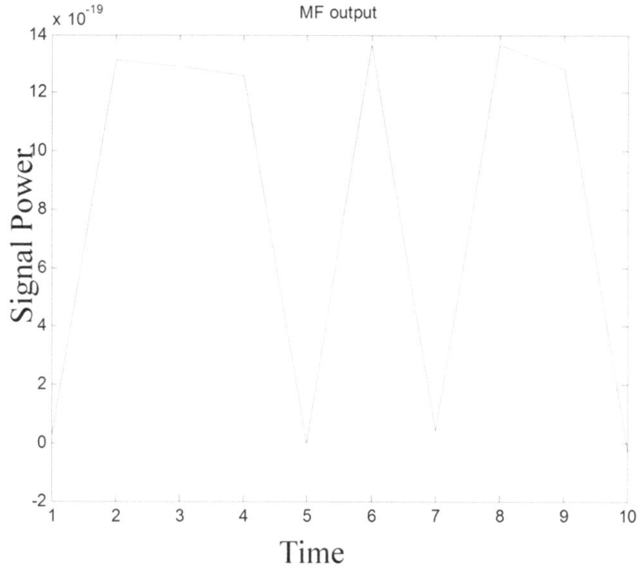

Figure 21.
Signal at the output of MF.

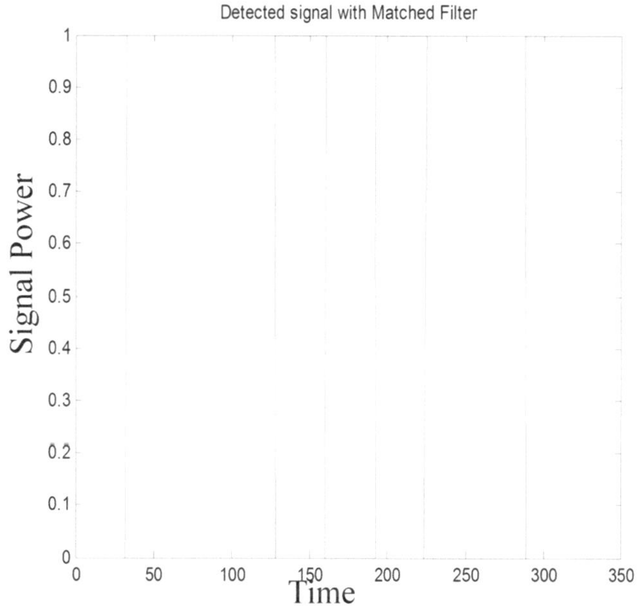

Figure 22.
Detected signal with MF.

10. BER variation with respect to SNR

Figures 23–31 represent the BER ~ SNR curves of OOK, BPSK and DPSK systems at data rate of 1 Gbps, 10 Gbps and 20 Gbps respectively as obtained through MATLAB simulations. Each of these contains a pair of curves, one of which indicates the system performance with MF and the other without the MF.

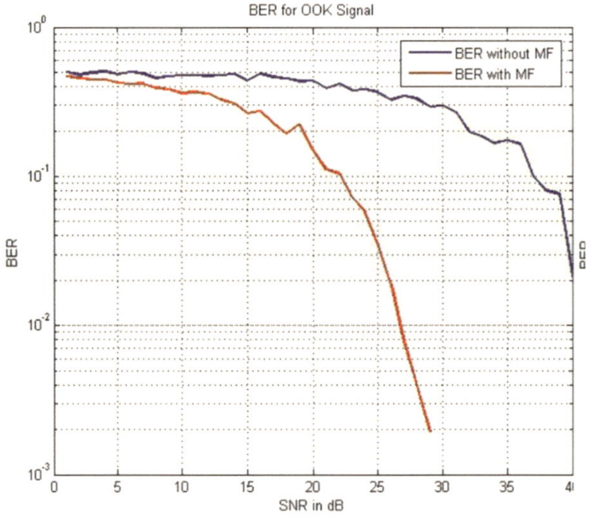

Figure 23.
BER curve for OOK at 1 Gbps.

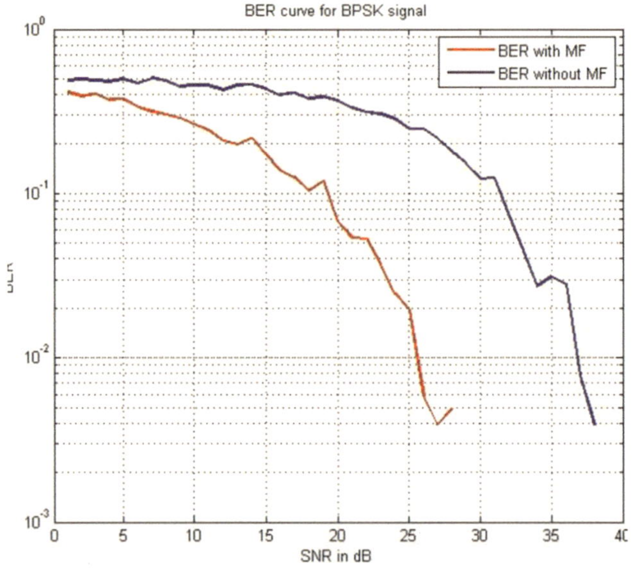

Figure 24.
BER curve for BPSK at 1 Gbps.

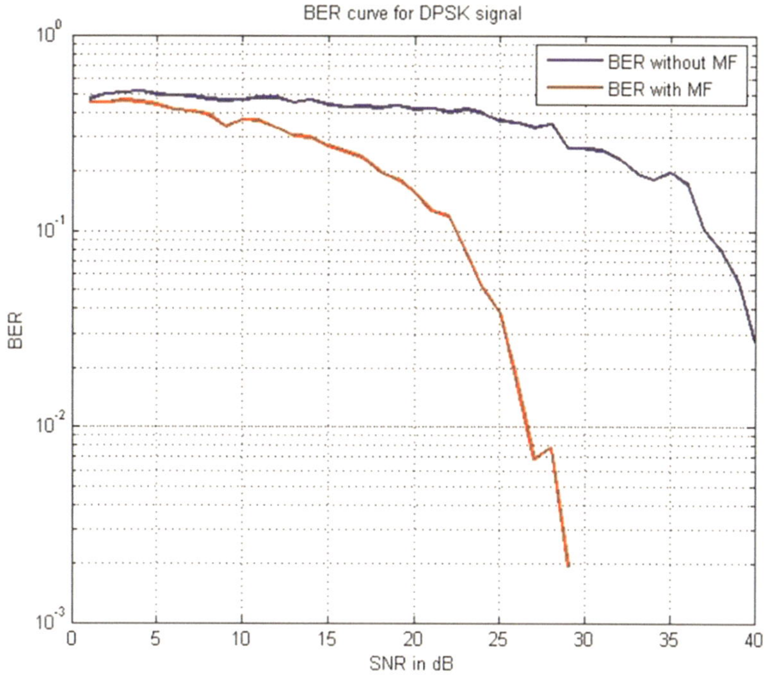

Figure 25.
BER curve for DPSK at 1 Gbps.

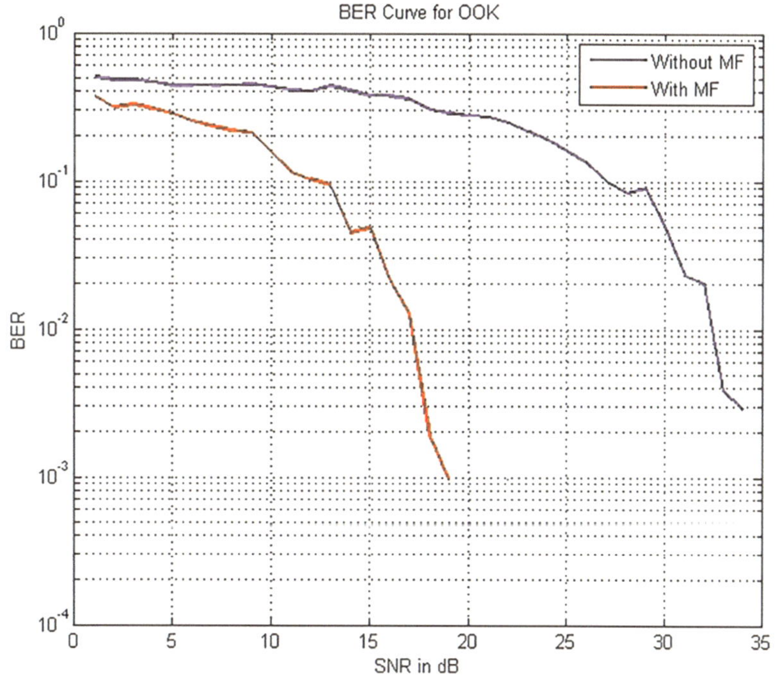

Figure 26.
BER curve for OOK at 10 Gbps.

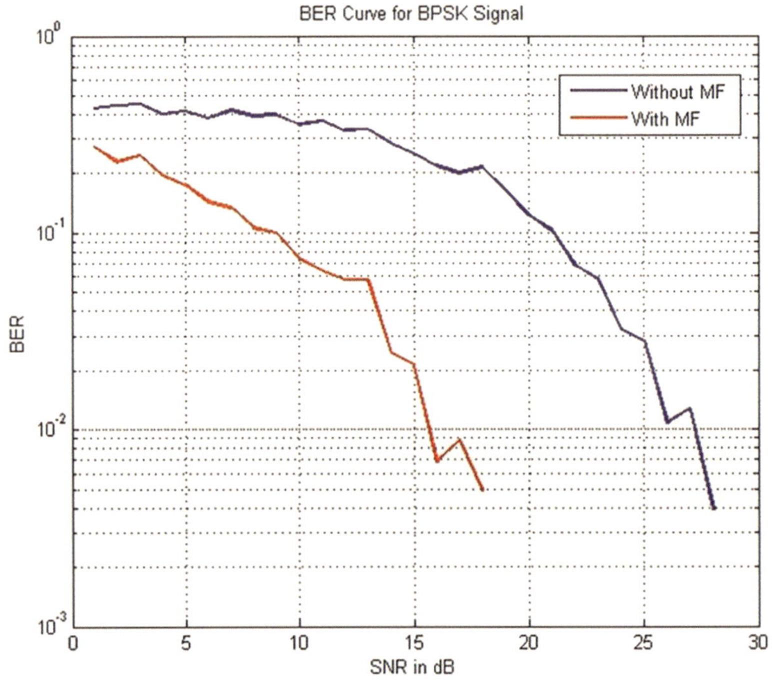

Figure 27.
BER curve for BPSK at 10 Gbps.

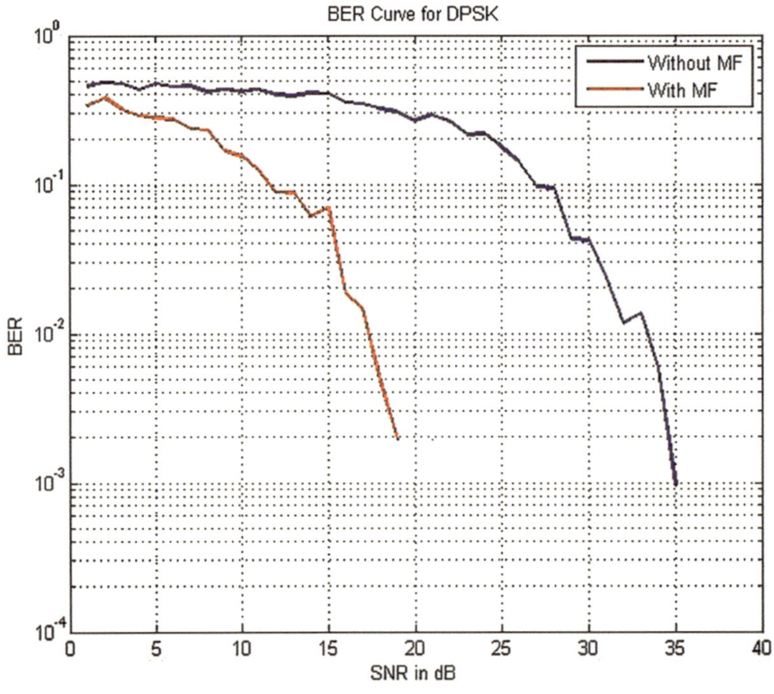

Figure 28.
BER curve for DPSK at 10 Gbps.

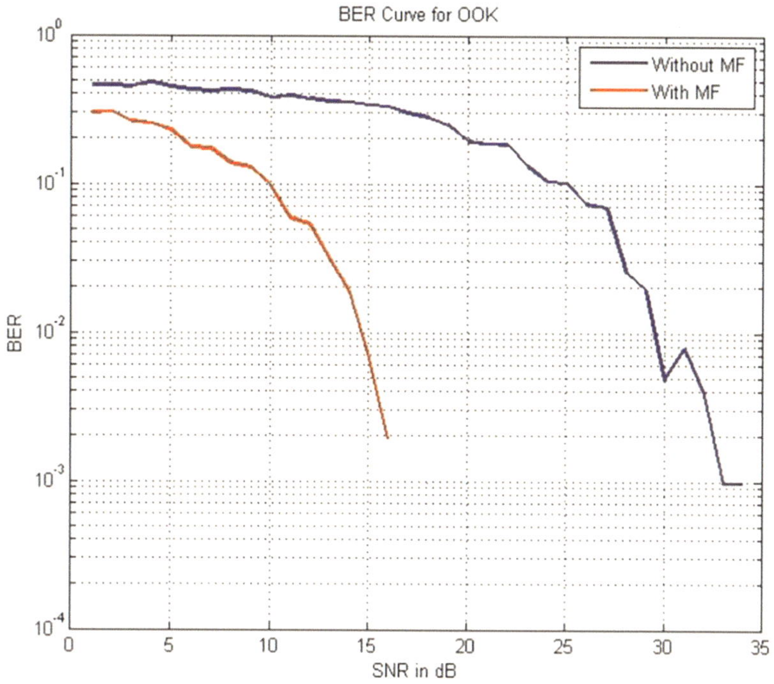

Figure 29.
BER curve for OOK at 20 Gbps.

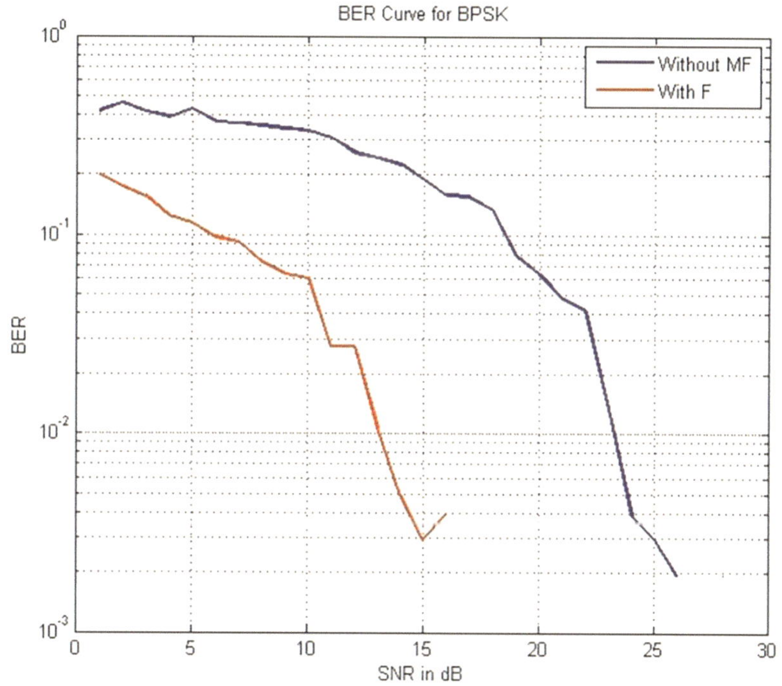

Figure 30.
BER curve for BPSK at 20 Gbps.

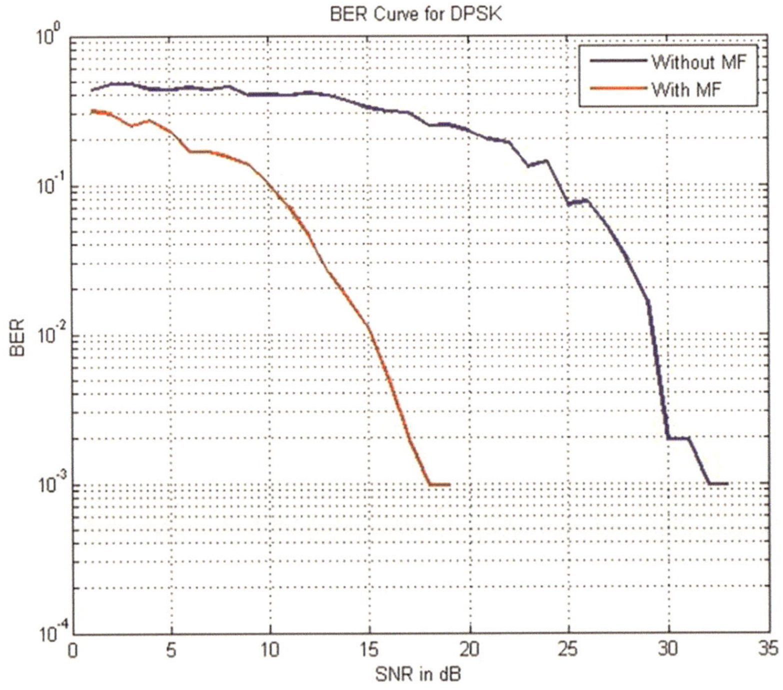

Figure 31.
BER curve for DPSK at 20 Gbps.

The purpose of taking three different data bits is to verify whether the MF is able to yield an acceptable signal quality in terms of SNR. By observing these figures it can be clearly noticed that for each of the systems the BER reduces at a higher rate and falls down to a more lower value when the demodulation is done with matched filter as compared to that in case of demodulation done without matched filter. A BER comparison for all three modulation schemes at the three different data rates is shown in **Tables 6** and 7.

Modulation scheme	Bit rate (GBPS)	Demodulation without MF		Demodulation with MF	
		Minimum BER	Value of SNR(dB) required	Minimum BER	Value of SNR(dB) required
OOK	1	0.02	40	0.002	28
	10	0.004	34	0.001	18
	20	0.001	34	0.002	16
BPSK	1	0.003	36	0.003	26
	10	0.004	28	0.005	18
	20	0.002	26	0.003	15
DPSK	1	0.03	40	0.001	28
	10	0.001	35	0.002	18
	20	0.001	32	0.001	18

Table 7.
Table of comparison for the system with and without MF.

The results obtained from **Table 6** reveal that in each case, between the minimum BER achieved through the use of MF and without using the MF, there is observed a minimum of 10 dB difference of SNR. With increasing data rate the there is not a significant variation in the minimum achievable BER.

11. Conclusion

Here the results reveal that with an increasing data rate the Q-factor goes on decreasing while BER increases. So the data rate may be a limitation for the transmitters. Making a comparison of the three modulation techniques, DPSK is found to have the best Q-factor in all the data rates taken in analysis. As noted in **Table 3**, the Q-factor in all cases is falling down with increasing data rate but the falling rate is lower in case of DPSK system as compared to the others. The Q-factor value at the worst scenario, that is at 40GBPS data rate falls to 2.24 and 4.03 in case OOK and BPSK system, while it is 5.63 in case of DPSK system. The detected output with using matched filter has been compared with the detected signal without using matched filter for three modulation schemes at three different data rates to assess their suitability over an OWC. The signal plots give information regarding the mismatch between the transmitted signal and detected signal without using MF whereas the MF insertion gives an acceptable signal quality. Each of the BER versus SNR plots also include two curves which show that the signal detected without MF needs a higher value of SNR to achieve a certain level of BER that can be achieved at a 10 dB (approximately) lesser value of SNR by a system having MF. The MF is not required in case if signal impairment is due to AWGN only. But practically, in OWC, the ambient light interference is a serious issue to be taken into consideration, because this effect cannot be avoided. This necessitates signal recovery through an MF.

Author details

Aruna Tripathy and Aishwarya Dash*
Odisha University of Technology and Research, Bhubaneswar, India

*Address all correspondence to: aishdash159@gmail.com

References

[1] Ghassemlooy Z, Popoola W, Rajbhandari S. Optical Wireless Communications Systems and Modeling with MATLAB. Boca Raton, London, New York: CRC Press; 2013

[2] Fritz RG, Bapst U.Wireless in-house data communication via diffuse infrared radiation. Proceedings of the IEEE. **67** (11):1474-1486

[3] Reddy GDK, Giri RK. Performance analysis of indoor optical wireless communication with and without FLI effect using various modulation schemes. In: Proceedings of the 2nd IEEE International Conference on Inventive Communication and Computational Technologies (ICICCT 2018). Piscataway, New Jersey, USA. 2018. pp. 574-577

[4] Kaur R, Kaur H. 'Comparative analysis of chirped, AMI and DPSK modulation techniques in IS-OWC system. Optik. 2018;**154**:755-762

[5] Pradhan S, Sahu PK, Giri RK, Patnaik B. Inter-Satellite Optical Wireless Communication System Design Using Diversity Techniques, International Conf. on Microwave, Optical and Communication Engineering2015. pp. 250-253

[6] He J, Norwood RA, Pearce MB, Djordjevic IB, Cvjetic M, Subramaniam S, et al. A survey on recent advances in optical communications. Computers and Electrical Engineering. 2014;**40**:216-240

[7] Kurniawan P, Sujatmoko K, Pamukti B. Performance of OOK-RZ and NRZ modulation techniques in various receiver positions for Li-Fi. In: International Conference on Signals and Systems (ICSigSys). New York: IEEE; 2019

[8] GPON-FTTX using NRZ and RZ modulation formats. Journal of Engineering and Applied Sciences. 2019; **14**(21):7951-7959

[9] Hadisiswoyo MR, Samijayani ON, Syahriar A, Arifianto I. Simulation of comparison from NRZ and RZ pulse in free space optics with different weather conditions. Universal Journal of Electrical and Electronic Engineering. 2019;**6**(3):151-158

[10] Groumas P, Katopodis V, Kouloumentas C, Bougioukos M, Hercules AK. All-optical RZ-to-NRZ conversion of advanced modulated signals. IEEE Photonics Technology Letters. 2012;**24**(3):179-181

[11] Klekamp A, Idler W, Dischler R. Comparison of FSK by directly modulated DFB laser with DPSK, NRZ and RZ modulation formats at 10 Gb/s. In: Researchgate, Conference Paper. Berlin, Germany. 2003

[12] Li Z, Dong Y, Lu C, Wen YJ, Wang Y, Hu W, et al. Comparison of cross-gain modulation effect of Manchester-Duobinary, RZ-DPSK, NRZ-DPSK, RZ, and NRZ modulation formats in SOAs. Piscataway, New Jersey, USA: IEEE Photonics Technology Letters. 2006;**24**:179-115

[13] Ohm M, Freckmann T. Comparison of different DQPSK transmitters with NRZ and RZ impulse shaping. In: IEEE/ LEOS Workshop on Advanced Modulation Formats. San Francisco, USA, CA; 2004

[14] Choudhury MZ, Jang TH. A comparative survey of optical wireless technologies: Architecture and applications. IEEE Access. 2018;**2018**: 9819-9840

[15] Binh LN. Digital Optical Communications. Boca Raton, London, New York: CRC Press; 2012

[16] Binh LN. Optical Fiber Communication Systems with MATLAB and Simulink Models. 2nd ed. Boca Raton, London, New York: CRC Press; 2015

[17] Lee S. Reducing the effects of ambient noise light in an indoor optical wireless system using polarizers. Microwave and Optical Technology Letters. 2004;**40**:228-230

[18] Hayes AR, Ghassemlooy Z, Seed NL, McLaughlin R. Baseline-wander effectson systems employing digital pulse-interval modulation. IEEE Proceedings—Optoelectronics. 2000; **147**:295-300

[19] Narasimhan R, Audeh MD, Kahn JM. Effect of electronic-ballast fluorescent lighting on wireless infrared links. IEEE Proceedings—Optoelectronics. 1996;**143**: 347-354

[20] Moreira AJC, Valadas RT, Duarte AMO. Optical interference produced by artificial light. Wireless Networks. 1997;**3**:131-140

[21] Turin GL. An introduction to matched filters. IRE Transactions on Information Theory. 1960;**6**(3):311-329. DOI: 10.1109/TIT.1960.1057571

Chapter 4

The Effect of Polymerization of 2.7-Divinylcarbazole-Benzo-Bis-Thiadiazole on Optical Fiber Properties

Mohamed Jabha, Abdellah El Alaoui, Abdellah Jarid and El Houssine Mabrouk

Abstract

The interest of polymer optical fibers (POF) lies in their low cost compared to silica fibers and in their ease of implementation, i.e. robustness, flexibility, low weight and easier connectivity. The first generation of polymer fibers are of the index jump type and are composed of polymethyl methacrylate (PMMA) for the core and a fluorinated polymer for the cladding. The significant attenuation of OPFs in the red and near IR is due to the harmonics of the different vibrational modes of the C–H bonds. The improvement of this parameter requires a shift in the transmission of the polymer towards longer wavelengths. As in the case of inorganic glasses, this requires the development of materials with low fundamental frequency of vibration. The development of graded index structures also allows limiting the modal dispersion inherent to the multi-mode character of POFs. And before the use of certain materials in the electronic fields a study of certain properties was carried out by the DFT method in order to propose the polymers based on carbazole. This study was carried out by the DFT–B3LYP method as functional with the 6-31G (d, p) atomic base to optimize all systems, from monomer to pentamer.

Keywords: carbazole, DFT, POF, oraganic technology, optical fiber properties

1. Introduction

The carbazole [1] drifts turns out to be a particularly well adapted material here, since it can be while keeping its optical and thermomechanical qualities exceptional. However, the advantage of polymers for planar technology is their good transparency over short distances, associated with an easier shaping than inorganic glasses.

2,7-Carbazole derivatives [2–4] are organic materials with the properties of a semiconductor. These oligomers are known for their high stability due to the presence of nitrogen atoms and exhibit important physicochemical and optoelectronic properties. The addition of the methoxy-benzyl group to the nitrogen atom increases the solubility of these oligomers and facilitates their synthesis. During the last decade, conjugated donor-acceptor (D − A) copolymers have been the subject of numerous studies [2]. Among the main advantages of these hetero-junction materials, they have high flexibility and low weight [3–5]. The objective of this work is to show the interest of π-conjugated polymers by focusing on the electronic structure, the factors influencing the gap energy and the substitution effect of these polymers. We are interested in increasing the efficiency of the organic photovoltaic cell, by seeking to decrease the gap energy (HOMO-LUMO), in characterizing these compounds in terms of geometric and electronic structures to ensure good absorption of radiation and facilitate the charge transfer between the different compounds of our copolymer. This study was carried out on copolymers based on 2,7-divinylcabazole (PCrV) and benzo-bis-thiadiazole (BBT) (**Figure 1**). We were interested in the effect, on the structural, optoelectronic and photovoltaic properties of the addition of the méthoxy-benzyl group on the N atom of carbazole. This study was carried out by the DFT–B3LYP method as functional with the 6-31G (d, p) [6–12] atomic base to optimize all systems, from monomer to pentamer. The structural and electronic properties have been demonstrated following this optimization, while the optical properties are obtained by the TD-DFT method at the level of the WB97XD functional and with the 6-31G base [13–24].

The LUMO level of the N-substituted copolymer is more stable than that of unsubstituted copolymer. This shows that the substitution of H by the méthoxy-benzyl group allows to decrease the energy gap of these compounds [25–31].

2. Structural properties

The structures of the Donor–Acceptor (DA) [32] copolymers based on 2,7-divinyl-carbazole and benzo-bis-thiadiazole (PCrV-BBT) have been optimized with the base B3LYP/6-31G (d, p). The geometric parameters (bond lengths d_i, angles A_i and

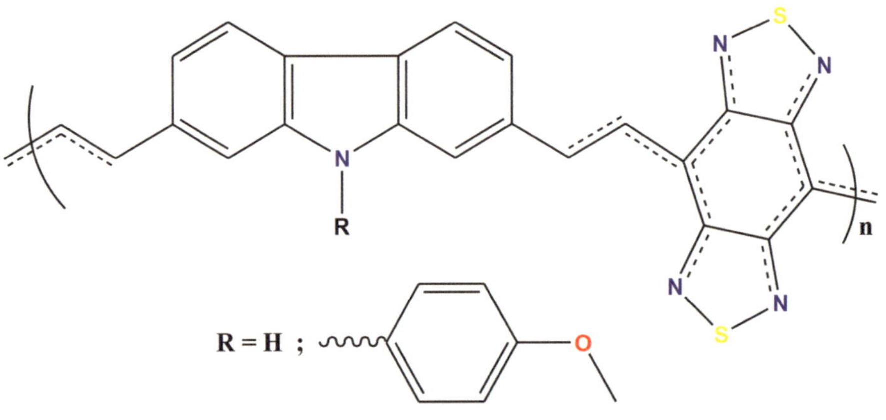

Figure 1.
The structures of the copolymers studied of 2, 7-divinyl-carbazole and Benzo-bis-thiadiazole (PCrV-BBT); N Substituted and N not Substituted.

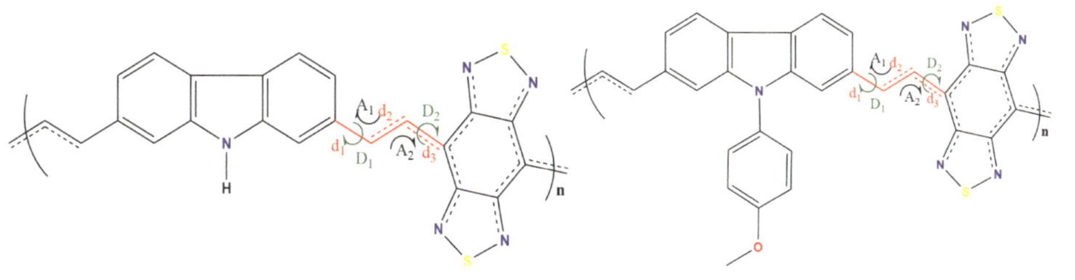

Figure 2.
The structures of the copolymers studied (of 2,7-divinyl-carbazole and benzo-bis-thiadiazole (PCrV-BBT; NH)), and PCrV-BBT; N-Sub (with the addition of methoxy - benzyl on the N atom of carbazole).

dihedral angles D_i) (**Figure 2**) of the most stable conformations obtained after the total optimization of the systems studied are presented in **Table 1**.

The values of the inter-cyclic bonds (d_i), the values of the normal angles (A_i) and the values of the dihedral angles (D_i) of the copolymers based on PCrV-BBT with the substituted carbazole nitrogen atom (PCrV-BBT; N-sub) are generally very similar to those of the copolymers based on PCrV-BBT unsubstituted in N carbazole (PCrV-BBT; NH) (**Table 1**).

The lengths of the inter-cyclic bonds (d_i, i = 1–3) of all the copolymers remain almost constant going from monomer to pentamer while maintaining the double and single bond alternation, i.e. an increase of the length of the conjugation with the extension of the carbon chain. Thus, the lengths of the bonds d_1 and d_3 have values close to the length of the single bond C–C (1.465 Å) and the value of R2 is close to that of the length of double bond C = C (1.361 Å) [33–35].

From the values mentioned in **Table 1**, it is observed that the values of the dihedral angles of the systems studied are generally porches of 180°. This indicates that the two benzo-bis-thiadiazole and 2,7-vinyl-carbazole moieties are found in the same plane in all systems (from n = 1 through n = 5). On the other hand, the pentamer has a planar structure with dihedral angle values very close to 180°. This flatness, confirmed by the values of the normal angles (A_i; i = 1, 2) close to 120°, reflects SP2 hybridization.

These results show a good improvement in flatness and an increase in electron conjugation π. This facilitates the possibility of delocalization of electrons along the molecular skeleton.

3. Electronic properties

The study of electronic properties mainly concerns the determination of energy levels HOMO, LUMO and the band gap (energy gap). This determination makes it possible to validate the candidacy of the molecules studied for use in the electronic field. The calculation of the HOMO, LUMO and Energy Gap levels was carried out after the optimization of the systems by the B3LYP/6-31 g (d, p) method.

From **Table 2**, we note that the gap values decrease going from monomer to pentamer. We also note that the addition of the methoxy-benzyl group on the nitrogen atom of the carbazole causes the gap to decrease by an average of 0.02 eV. This decrease due to a good stability of the energy levels is clearly illustrated in **Figure 3**. This decrease in the gap is in agreement with the improvement in the flatness of these systems.

Parameters		PCrV-BBT; N-H)n					PCrV-BTT; N-sub)n				
		n = 1	n = 2	n = 3	n = 4	n = 5	n = 1	n = 2	n = 3	n = 4	n = 5
Inter-cyclic distance (Å)	d_1	1.455	1.454	1.454	1.454	1.454	1.455	1.454	1.454	1.453	1.453
	d_2	1.360	1.362	1.36	1.362	1.362	1.360	1.360	1.362	1.363	1.363
	d_3	1.440	1.436	1.436	1.436	1.436	1.439	1.435	1.435	1.435	1.435
Normal angles (°)	A_1	117.65	117.74	117.81	117.75	117.74	117.7	117.68	117.68	117.63	117.71
	A_2	127.15	127.65	127.68	127.66	127.63	127.36	127.68	127.67	127.59	127.49
Dihedral angles (°)	D_1	180.01	179.99	179.93	179.96	180.04	179.72	179.52	179.23	179.56	179.92
	D_2	179.99	180.0	179.98	179.98	180.02	179.89	179.88	179.95	179.86	179.80

Table 1.
The structural parameters of the copolymers; d_i: inter-cyclic distances, A_i: normal angles, D_i: dihedral angles.

Polymers	PCrV-BBT; N-H			PCrV-BBT; N-Sub		
	HOMO (eV)	LUMO (eV)	GAP Energy (eV)	HOMO (eV)	LUMO (eV)	GAP Energy (eV)
n = 1	−5.16	−3.40	1.76	−5.09	−3.34	1.75
n = 2	−4.77	−3.49	1.28	−4.68	−3.42	1.26
n = 3	−4.66	−3.53	1.13	−4.58	−3.46	1.12
n = 4	−4.61	−3.54	1.07	−4.53	−3.47	1.06
n = 5	−4.59	−3.56	1.03	−4.49	−3.48	1.01

Table 2.
The values of the energies of the HOMO and LUMO levels and of the energy gap of the copolymers.

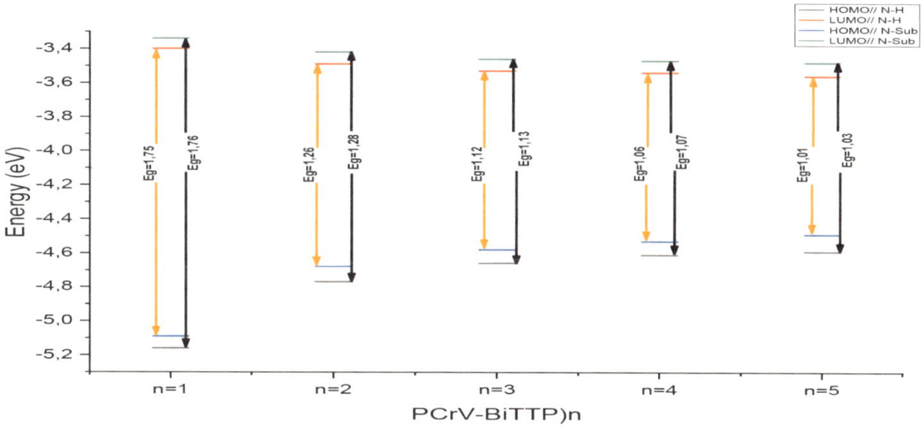

Figure 3.
Illustration of the HOMO, LUMO, and Eg levels of the monomer to the pentamer obtained by B3LYP/6-31G (d, p).

The change from monomer to pentamer shows a general reduction in the gap with a strong stability of the LUMO level compared to the HOMO level, which seems a little destabilized (see **Figure 3**). We also observe that the LUMO levels of PCrV-BBT-N-substituted are more stable than those of unsubstituted PCrV-BBT-N-H. This clearly shows that the substitution of the H atom by the methoxy-benzyl group allows a reduction in the gap of these copolymers.

Concerning the electronic densities of the HOMO and LUMO border molecular orbitals of the copolymers studied, they are presented in **Figure 4**.

This figure shows that the electron density at the level of HOMO orbitals is distributed along the conjugation chain. In contrast, at LUMO orbitals, the electron cloud is mainly located on the electron acceptor unit (BBT). This phenomenon is due to the Puch-Pull properties of type polymers (D–A). It should be noted that the addition of the methoxy-benzyl group to the nitrogen of the carbazole promoting the flatness of these systems, allows the total displacement of the electron density towards the acceptor unit (**Figure 5**) and plays an important role at the level of the solubility and stability of the systems [4, 36]. It is also important to note that the transfer of electrons from the HOMO orbital to the LUMO orbital in the N-substituted copolymer is more efficient compared to other N-unsubstituted copolymers and therefore results in energy low gap.

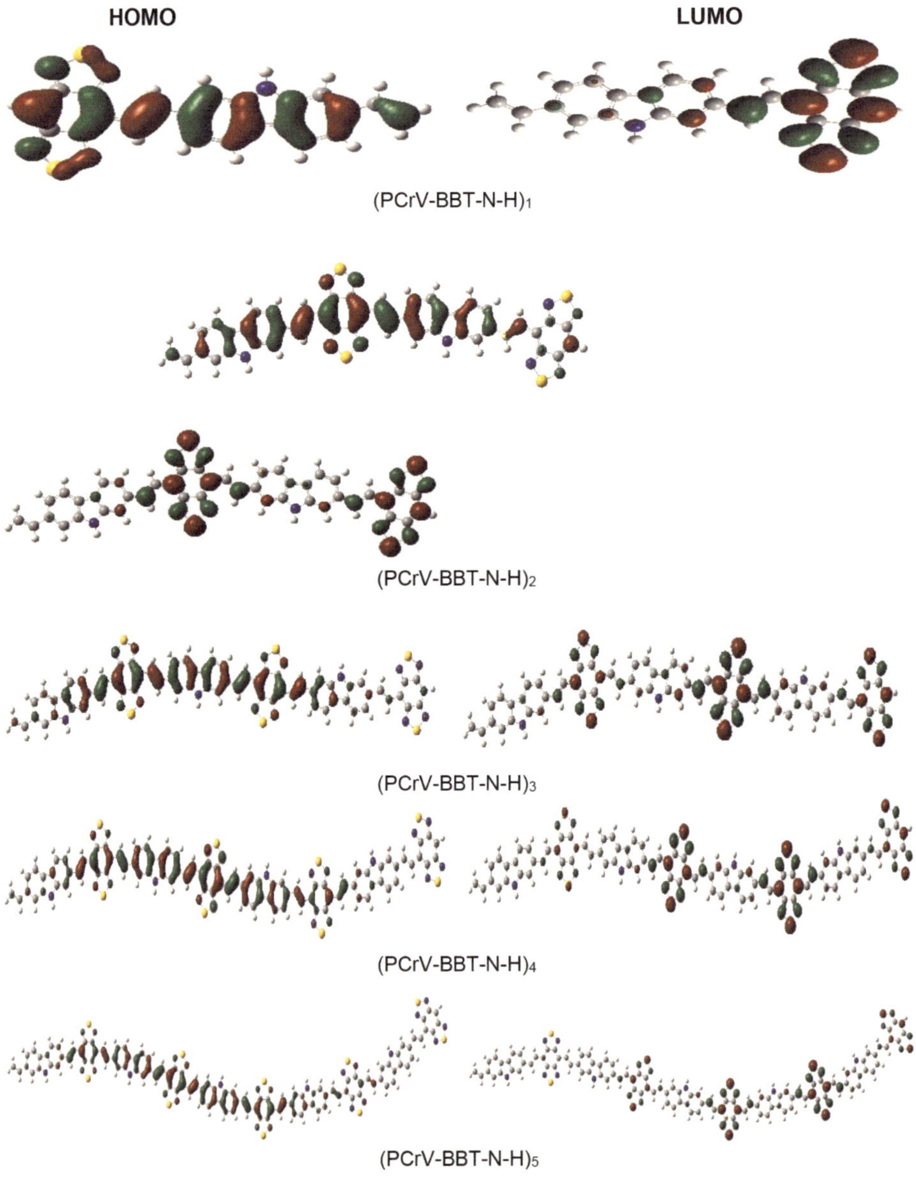

4. Optical properties and electronic transitions

UV–visible absorption spectra, wavelengths and excitation energies of the copolymers studied were calculated using the TD-DFT-WB97XD/6-31G method, from the optimized geometries of the copolymers. The corresponding values of excitation energies (Eex), wavelengths (λmax), and light harvesting efficiency (LHE) [expressed by Eq. (1)] [37], are gathered in **Table 3**.

To understand the evolution of the short-circuit current density (Jsc), we studied the LHE energy which depends on Jsc according to Eq. (2) [38]. We notice that large values of LHE lead to high values of Jsc. This therefore improves the efficiency of electronic devices.

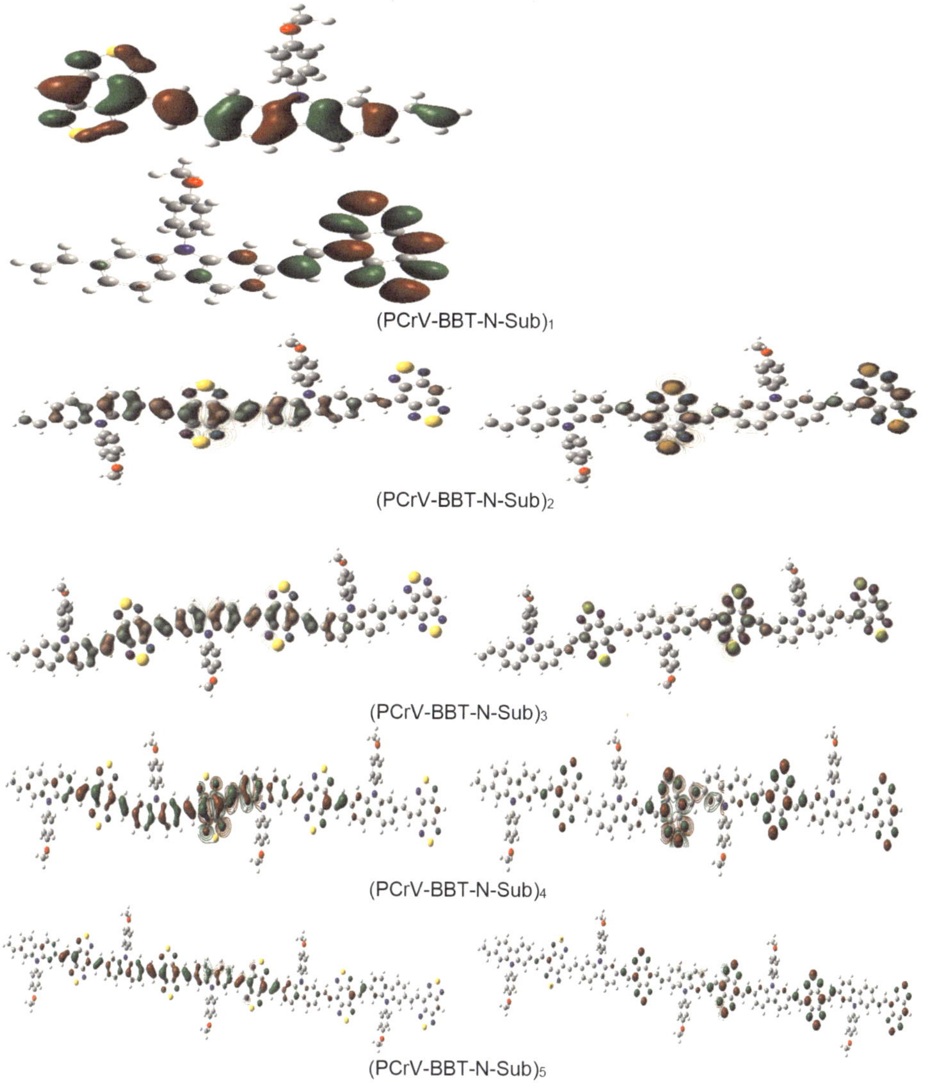

Figure 4.
Illustration of the border orbitals HOMO, LUMO of the copolymers without group (PCrV-BBT; N-H), and with methoxy-benzyl group (PCrV-BBT; N-Sub).

$$\mathrm{LHE} = 1\text{–}10 - \mathrm{f} \tag{1}$$

$$\mathrm{Jsc} = \int \lambda\ \mathrm{LHE}(\lambda)\ \phi(\mathrm{inject})\ \eta(\mathrm{collect})\ d\lambda \tag{2}$$

where: ϕ(inject), Efficiency of electron injection; η(collect), The efficiency of the collection of charges; f, oscillation force.

From the data in **Table 3**, we have three types of same-spin (S→S) excitation from the lowest occupied molecular orbitals and the highest void levels. In regards to λ_{max}, this increases going from the monomer to the pentamer. In addition, the addition of the methoxy-benzyl group contributes to increasing the values of maximum absorption lengths. This indicates that these copolymers are good candidates for absorbing the maximum of incident light radiation and therefore further increasing the

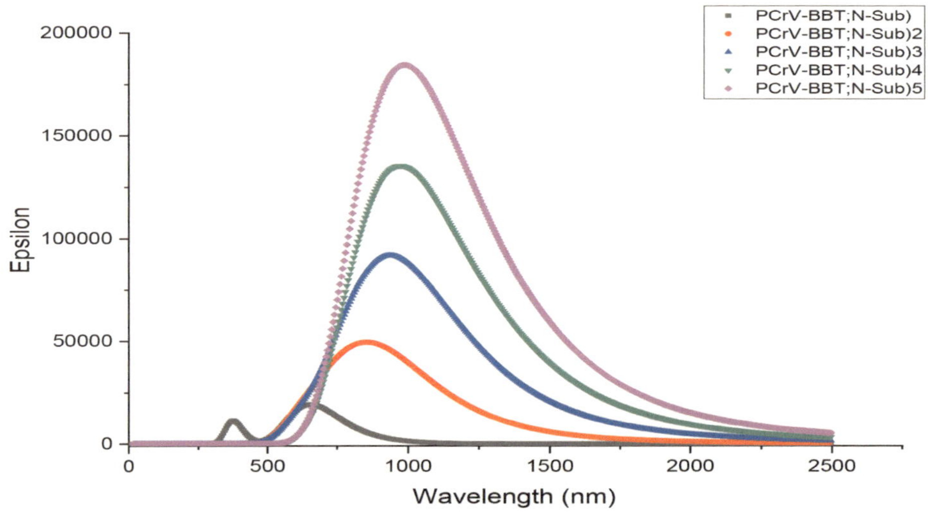

Figure 5.
Data of wavelength according to absorbance of copolymers with the methoxy-benzyl group on N of carbazole obtained at DFT-WB97XD/6-31G.

photoelectric conversion efficiency of the corresponding solar cell from this type of material (**Figure 6**).

These absorption spectra show an increase in the magnitude of epsilon directly related to the absorbance of light. This implies very good absorption of light radiation going from the monomer to the pentamer. The value of λ_{max} increases until it reaches the far IR zone. In other words, it only takes a gentle incident light radiation to have a very good excitation of the electrons. We also find that the width of the excitation plug increases when the methoxy-benzyl group is substituted on the N atom of the carbazole. This is favorable to good absorption of the light ray.

5. NBO analysis and charge transfer

From the optimized geometries of the copolymers, we fragmented the systems into donor fragments and other acceptors. Then we carried out the energy calculation by the DFT-B3LYP/6-31G (d, p) method.

NBO analysis [28–30] is an efficient method to study the interaction between intra and intermolecular bonds. It also allows the study of charge transfer in the molecular system.

Figure 7 above represents the different fragments of the two types of copolymers studied, we chose to fragment the structure according to the copolymer compositions, then we performed the NBO analysis, the results obtained are grouped together in **Table 4**.

NBO analysis of the various fragments of the copolymer highlights the role of the interaction of intermolecular orbitals and charge transfer in the system. It is carried out by taking into account all the possible interactions between filled donors and empty acceptors and by estimating their energetic importance by the theory of second order disturbances.

	Type of excitement	Eex (eV)	λ (nm)	LHF	Attributions
PCrV-BBT; N-Sub	S0 ➔ S1	1.8983	653.13	0.6651	H-2 ➔ L (11%); H ➔ L (88%)
	S0 ➔ S2	3.1570	392.73	0.0355	H-5 ➔ L1 5(3%); H-1 ➔ L (86%); H ➔ L + 1 (6%)
	S0 ➔ S3	3.3028	375.39	0.4563	H-8 ➔ L (2%); H-4 ➔ L (24%); H-2 ➔ L (59%); H-1 ➔ L (2%); H ➔ L (7%); H ➔ L + 1 (3%)
PCrV-BBT; N-Sub)2	S0 ➔ S1	1.3976	887.09	0.9244	H-1 ➔ L + 1(4%); H- ➔ L (77%); H ➔ L + 1 (14%)
	S0 ➔ S2	1.8953	654.16	0.6036	H-5 ➔ L + 1 (3%); H-4 ➔ L + 1 (4%); H-1 ➔ L (22%); H-1 ➔ L + 1(49%); H ➔ L + 1 (19%)
	S0 ➔ S3	2.8687	432.20	0.0059	H-8 ➔ L (6%); H-8 ➔ L + 1 (2%); H-5 ➔ L (9%); H-4 ➔ L (24%); H-4 ➔ L + 1 (6%); H-3 ➔ L (2%); H-1 ➔ L (32%); H-1 ➔ L + 1 (9%)
PCrV-BBT; N-Sub)3	S0 ➔ S1	1.3107	945.93	0.9938	H-2 ➔ L + 2 (3%); H-1 ➔ L (3%); H-1 ➔ L + 1 (21%); H-1 ➔ L + 2 (8%); H ➔ L (54%); H ➔ L + 1(6%)
	S0 ➔ S2	1.4693	843.82	0.0243	; H-1 ➔ L (34%); H-1 ➔ L + 2 (5%); H ➔ L + 1 (38%) H ➔ L + 2 (14%)
	S0 ➔ S3	1.8973	653.49	0.6033	H-7 ➔ L + 2 (2%); H-2 ➔ L (11%); H-2 ➔ L + 1 (19%); H-2 ➔ L + 2 (38%); H-1 ➔ L + 1 (3%); H-1 ➔ L + 2 (12%); H ➔ L + 2 (7%)
PCrV-BBT; N-Sub)4	S0 ➔ S1	1.2684	977.48	0.9994	H-2 ➔ L + 2 (14%);; H-2 ➔ L + 3 (6%); H-1 ➔ L + 1 (17%); H-1 ➔ L + 2 (3%); H ➔ L (44%); H ➔ L + 1 (4%)
	S0 ➔ S2	1.4041	883.05	0.0212	H-2 ➔ L + 1 (12%); H-2 ➔ L + 2 (2%); H-1 ➔ L (27%); H-1 ➔ L + 2 (10%); H-1 ➔ L + 3 (9%); H ➔ L + 1 (26%); H ➔ L + 2 (5%);
	S0 ➔ S3	1.4957	828.95	0.2504	H-2 ➔ L (25%); H-2 ➔ L + 1 (3%); H-1 ➔ L + 1 (19%); H-1 ➔ L + 2 (3%); H ➔ L + 2 (26%); H ➔ L + 3 (13%)
PCrV-BBT; N-Sub)5	S0 ➔ S1	1.2492	992.49	0.9999	H-3 ➔ L + 2 (2%); H-3 ➔ L + 3 (6%); H-3 ➔ L + 4 (8%); H-2 ➔ L + 1 (3%); H-1 ➔ L + 2 (9%); H-2 ➔ L + 3 (3%); H-1 ➔ L (5%); H-1 ➔ L + 1 (12%); H-1 ➔ L + 2 (5%); H ➔ L(32%); H ➔ L + 1 (8%);
	S0 ➔ S2	1.3543	915.48	0.0076	H-3 ➔ L + 1 (2%); H-3 ➔ L + 2 (6%); H-3 ➔ L + 3 (3%); H-2 ➔ L (3%); H-2 ➔ L + 1 (7%); H-2 ➔ L + 3 (4%); H-2 ➔ L + 4 (8%); H-1 ➔ L (23%); H-1 ➔ L + 2 (6%); H ➔ L + 3 (4%); H ➔ L + 1 (20%); H ➔ L + 2 (5%);
	S0 ➔ S3	1.4491	855.60	0.3458	H-3 ➔ L + 1 (7%); H-3 ➔ L + 2 (4%); H-2 ➔ L (18%); H-2 ➔ L + 2 (4%); H-2 ➔ L + 3 (3%); H-1 ➔ L + 1 (13%); H-1 ➔ L + 3 (5%); H-1 ➔ L + 4 (11%); H ➔ L + 2 (18%); H ➔ L + 3 (5%);
PCrV-BBT; N-H	S0 ➔ S1	1.9137	647.86	0.6744	H-2 ➔ L (11%); H ➔ L (88%);
	S0 ➔ S2	3.3145	374.07	0.4419	H-6 ➔ L (2%); H-3 ➔ L (23%); H-2 ➔ L (55%); H-1 ➔ L (8%); H ➔ L (7%);
	S0 ➔ S3	3.3758	367.28	0.0744	H-4 ➔ L (5%); H-2 ➔ L (5%); H-1 ➔ L (79%); H-1 ➔ L + 1 (6%);

	Type of excitement	Eex (eV)	λ (nm)	LHF	Attributions
PCrV-BBT; N-H)2	S0 ➔ S1	1.4144	876.58	0.9259	H-1 ➔ L + 1(4%); H ➔ L (78%); H ➔ L + 1 (13%);
	S0 ➔ S2	1.9089	649.50	0.5977	H-5 ➔ L + 1 (3%); H-2 ➔ L + 1 (4%); H-1 ➔ L (22%); H-1 ➔ L + 1 (49%); H ➔ L + 1 (19%);
	S0 ➔ S3	2.8812	430.31	0.0341	H-6 ➔ L (6%); H-6 ➔ L + 1(2%); H-5 ➔ L (10%); H-2 ➔ L (25%); H-2 ➔ L + 1 (6%); H-1 ➔ L (33%); H-1 ➔ L + 1 (9%);
PCrV-BBT; N-H)3	S0 ➔ S1	1.3283	933.38	0.9943	H-2 ➔ L + 2 (3%); H-1 ➔ L (3%); H-1 ➔ L + 1 (22%); H-1 ➔ L + 2 (8%); H ➔ L (54%); H ➔ L + 1 (6%);
	S0 ➔ S2	1.4920	830.97	0.0561	H-1 ➔ L (35%); H-1 ➔ L + 1 (4%); H ➔ L + 1 (38%); H ➔ L + 2 (14%);
	S0 ➔ S3	1.9109	648.82	0.5951	H-7 ➔ L + 2 (2%); H-2 ➔ L (11%); H-2 ➔ L + 1 (19%); H-2 ➔ L + 2 (38%); H-1 ➔ L + 1 (3%); H-1 ➔ L + 2 (12%); H ➔ L + 2 (7%)
PCrV-BBT; N-H)4	S0 ➔ S1	1.2846	965.15	0.9994	H-2 ➔ L + 1 (3%); H-2 ➔ L + 2 (12%); H-2 ➔ L + 3 (7%); H-1 ➔ L (4%); H-1 ➔ L + 1 (15%); H-1 ➔ L + 2 (4%); H ➔ L (42%); H ➔ L + 1 (7%);
	S0 ➔ S2	1.4199	873.21	0.0043	H-2 ➔ L + 1 (10%); H-2 ➔ L + 2 (3%); H-1 ➔ L (26%); H-1 ➔ L + 2 (10%); H-1 ➔ L + 3(9%); H ➔ L + 1 (25%); H ➔ L + 2 (5%);
	S0 ➔ S3	1.5178	816.86	0.2333	H-2 ➔ L (24%); H-2 ➔ L + 1 (4%); H-1 ➔ L + 1 (18%); H-1 ➔ L + 2(3%); H ➔ L + 1 (2%); H ➔ L + 2 (26%); H ➔ L + 3 (13%);
PCrV-BBT; N-H)5	S0 ➔ S1	1.2600	983.96	0.9999	H-3 ➔ L + 2 (3%); H-3 ➔ L + 3 (8%); H-3 ➔ L + 4 (6%); H-2 ➔ L + 1 (4%); H-2 ➔ L + 2 (9%); H-2 ➔ L + 3 (3%); H-1 ➔ L (5%); H-1 ➔ L + 1 (11%); H-1 ➔ L + 2 (5%); H ➔ L (32%); H ➔ L + 1 (8%);
	S0 ➔ S2	1.3689	905.75	0.2034	H-3 ➔ L + 1 (2%); H-3 ➔ L + 2 (6%); H-3 ➔ L + 3 (2%); H-2 ➔ L (3%); H-2 ➔ L + 1 (7%); H-2 ➔ L + 3 (5%); H-2 ➔ L + 4 (7%); H-1 ➔ L (22%); H-1 ➔ L + 2 (6%); H-1 ➔ L + 3 (4%); H ➔ L + 1 (20%); H ➔ L + 2 (6%);
	S0 ➔ S3	1.4661	845.69	0.4410	H-3 ➔ L + 1 (7%); H-3 ➔ L + 2 (3%); H-2 ➔ L (18%); H-2 ➔ L + 2 (4%); H-2 ➔ L + 3 (2%); H-1 ➔ L + 1 (13%); H-1 ➔ L + 3 (6%); H-1 ➔ L + 4 (10%); H ➔ L + 1 (2%); H ➔ L + 2 (17%); H ➔ L + 3 (6%);

Table 3.
Types of excitation, excitation energies, wavelengths and LHE, obtained by the TD-DFT/WB97xd-6-31G method.

Each positively charged NBO represents a donor (i) and each negatively charged NBO is linked to an acceptor moiety (j). The stabilization energy E associated with electronic delocalization between donor and acceptor is estimated by Eq. (3) [39, 40].

$$E = qi\ (Fi.j)2/\varepsilon j - \varepsilon i \tag{3}$$

where: qi is the occupation of the i orbital; εj, εi are diagonal elements; Fi,j is the o-diagonal NBO Fock matrix element.

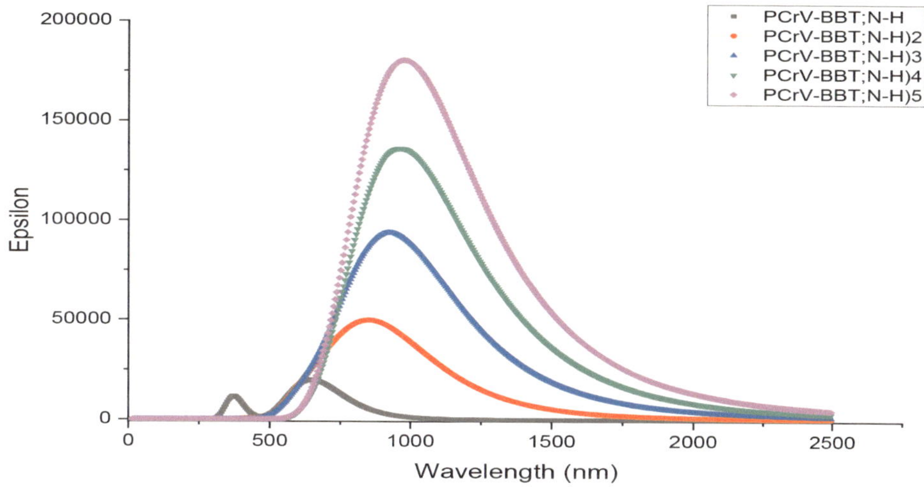

Figure 6.
Data of wavelength according to absorbance of copolymers with the H on N of carbazole obtained at DFT-WB97XD/6-31G.

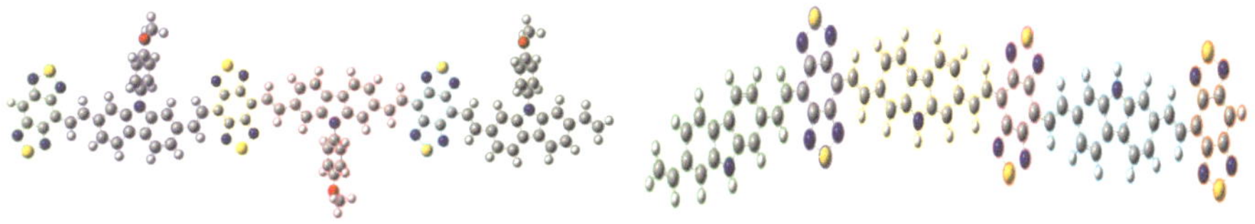

Figure 7.
Structures of the fragmented systems studied by NBO of copolymers without and with methoxy-benzyl group.

From **Table 4**, it is observed that the change from monomer to pentamer leads to the location of charges in the middle of the system; the charge values of the middle fragments are higher than those of the end fragments. The addition of the methoxy-benzyl group further increases the absolute value of charge compared to that of copolymers without a group (PCrV-BBT; N-H).

		PCrV-BBT; N-H		PCrV-BBT; N-Sub	
PCrV-BBT) n	Fragments	Charges	Types	Charges	Types
n = 1	Frag1	0.50	Donor	0.50	Donor
	Frag2	−0.50	Acceptor	−0.50	Acceptor
n = 2	Frag1	−0.15	Acceptor	0.13	Donor
	Frag2	0.13	Donor	−0.24	Acceptor
	Frag3	−0.23	Acceptor	0.26	Donor
	Frag4	0.25	Donor	−0.15	Acceptor
n = 3	Frag1	0.82	Donor	−0.15	Acceptor
	Frag2	0.13	Donor	0.26	Donor
	Frag3	−0.24	Acceptor	−0.24	Acceptor

		PCrV-BBT; N-H		PCrV-BBT; N-Sub	
PCrV-BBT) n	Fragments	Charges	Types	Charges	Types
	Frag4	0.24	Donor	0.25	Donor
	Frag5	−1.20	Acceptor	−0.24	Acceptor
	Frag6	0.25	Donor	0.13	Donor
n = 4	Frag1	0.24	Donor	0.19	Donor
	Frag2	0.13	Donor	−0.35	Acceptor
	Frag3	−0.24	Acceptor	0.36	Donor
	Frag4	−0.24	Acceptor	−0.36	Acceptor
	Frag5	0.24	Donor	0.37	Donor
	Frag6	−0.24	Acceptor	−0.35	Acceptor
	Frag7	0.25	Donor	0.38	Donor
	Frag8	−0.15	Acceptor	−0.25	Acceptor
n = 5	Frag1	0.13	Donor	−0.15	Acceptor
	Frag2	−0.14	Acceptor	0.26	Donor
	Frag3	0.25	Donor	−0.24	Acceptor
	Frag4	−0.24	Acceptor	0.25	Donor
	Frag5	0.24	Donor	−0.24	Acceptor
	Frag6	−0.24	Acceptor	0.25	Donor
	Frag7	0.24	Donor	−0.25	Acceptor
	Frag8	−0.24	Acceptor	0.24	Donor
	Frag9	0.24	Donor	−0.25	Acceptor
	Frag10	−0.24	Acceptor	0.13	Donor

Table 4.
The electronic charge and the type of fragments of the copolymers obtained by B3LYP/6-31G (d, p).

6. Conclusions

The conjugated copolymers based on 2,7-divinylcabazole (PCrV) and benzo-bis-thiadiazole (BBT) have been studied by suitable quantum methods. It emerges from this study that the electronic and optoelectronic properties are intimately linked to the molecular structure. In this sense, the electron donor and electron acceptor motif alternation approach is an effective strategy for modulating these properties.

The LUMO level of the N-substituted copolymer is more stable than that of unsubstituted copolymer. This shows that the substitution of H by the méthoxy-benzyl group allows to decrease the energy gap of these compounds.

The data obtained from the energy levels of the different molecules indicate that the LUMO and HOMO levels, as well as the values of the energy of gap Eg, vary according to the length of the carbon chain. The majority of the copolymers studied exhibit a high quality P-type semiconductor character. Electronic transitions in this type of material require low energy light radiation. For all the reasons mentioned above, these molecules are good candidates for a good application in electronic.

Author details

Mohamed Jabha[1,2]*, Abdellah El Alaoui[3], Abdellah Jarid[2] and El Houssine Mabrouk[1,4]

1 Faculty of Sciences and Technics, University of Moulay Ismail, Errachidia, Morocco

2 Faculty of Sciences Semlalia, University of Cadi Ayyad, Marrakech, Morocco

3 Faculty of Sciences, University of Moulay Ismail, Meknes, Morocco

4 Faculty of Sciences, Sidi Mohamed Ben Abdellah University, Fez, Morocco

*Address all correspondence to: m.jabha@edu.umi.ac.ma

References

[1] Leclerc N, Michaud A, Sirois K, Morin JF, Leclerc M. Advanced Functional Materials. 2006;**16**(13):1694. DOI: 10.1002/adfm.200600171

[2] Leliège A et al. Chemistry - A European Journal. 2013;**19**(30):9948. DOI: 10.1002/chem.201301054

[3] Aazou S et al. Journal of Optoelectronics and Advanced Materials. 2013;**13**(5–6):395. DOI: 10.1038/pj.2015.19

[4] Baek MJ, Lee S-H, Kim DH, Lee Y-S. Macromolecular Research. 2012;**20**(2):147

[5] Kim Y, Cho HH, Kim T, Liao K, Kim BJ. Polymer Journal. 2016;**48**(4): 517. DOI: 10.1038/pj.2016.22

[6] Hohenberg P, Kohn W. Physics Review. 1964;**136**:B864-BB71. DOI: 10.1103/PhysRev.136.B864

[7] Pople JA, Gill PMW, Johnson BG. Kohn-Sham density-functional theory within a finite basis set. Chemical Physics Letters. 1992;**199**(6):557. DOI: 10.1016/0009-2614(92)85009-Y

[8] Frisch MJ, Pople JA, Binkley JS. Self-consistent molecular orbital methods. The Journal of Chemical Physics. 1984; **80**:326. DOI: 10.1063/1.447079

[9] Stratmann RE, Burant JC, Scuseria GE, Frisch MJ. Improving harmonic vibrational frequencies calculation. The Journal of Chemical Physics. 1997;**106**:10175. DOI: 10.1063/1.474047

[10] Becke AD. Density-functional exchange-energy approximation with correct asymptotic behavior. Physical Review A. 1988;**38**(6):3098. DOI: 10.1103/physreva.38.3098

[11] Baker J, Andzelm J, Muir M, Taylor PR. Chemical Physics Letters. 1995;**237**:53. DOI: 10.1016/0009-2614(95)00299-J

[12] El Malki Z, Bouzzine SM, Bejjit L, Haddad M, Hamidi M, Bouachrine M, Density functional theory [B3LYP/6-311G(d,p)] study of a new copolymer based on carbazole and (3,4-Ethylenedioxythiophene) in their aromatic and polaronic states, Journal of Applied Polymer Science. 2011;**122**(5): 3351. https://doi.org/10.1002/app.34395

[13] Ganji MD, Tajbakhsh M, Kariminasab M, Alinezhad H. Tuning the LUMO level of organic photovoltaic solar cells by conjugately fusing graphene flake: A DFT-B3LYP study. Physica E: Low-dimensional Systems and Nanostructures. 2016;**81**:108. DOI: 10.1016/j.physe.2016.03.008

[14] Barone V. Chemical Physics Letters. 1994;**226**:392. DOI: 10.1016/0009-2614(94)00725-X

[15] Petersson GA, Al-Laham MA. A complete basis set model chemistry, II. Open-shell systems and the total energies. The Journal of Chemical Physics. 1991;**94**:6081. DOI: 10.1063/1.460447

[16] Petersson GA, Bennett A, Tensfeldt TG, Al-Laham MA, Shirley WA, Mantzaris J. A complete basis set model chemistry, I. The total energies of closed-shell atoms and hydrides of the first-row elements. Journal of Chemical Physics. 1988;**89**: 2193. DOI: 10.1063/1.455064

[17] Rassolov VA, Ratner MA, Pople JA, Redfern PC, Curtiss LA. 6-31G* basis set for third-row atoms. Journal of

Computational Chemistry. 2001;**22**(9): 976. DOI: 10.1002/jcc.1058urtiss

[18] Ditchfield R, Hehre WJ, Pople JA. Self-consistent molecular orbital methods. 9. Extended Gaussian-type basis for molecular-orbital studies of organic molecules. The Journal of Chemical Physics. 1971;**54**:724. DOI: 10.1063/1.1674902

[19] Cavillot V, Champagne B. Time-dependent density functional theory simulation of UV/visible absorption spectra of zirconocene catalysts. Chemical Physics Letters. 2002; **354**(5–6):449. DOI: 10.1016/S0009-2614(02)00161-6

[20] Jamorski JC, Lüthi HP. Time-dependent density-functional theory investigation of the formation of the charge transfer excited state for a series of aromatic donor–acceptor systems. Part I, The Journal of Chemical Physics. 2002;**117**(9):4146

[21] Adamo C, Scuseria GE, Barone V. Accurate excitation energies from time-dependent density functional theory: Assessing the PBE0 model. The Journal of Chemical Physics. 1999;**111**(7):2889. DOI: 10.1063/1.479571

[22] Chai JD, Head-Gordon M. Long-range corrected hybrid density functionals with damped atom–atom dispersion corrections. Physical Chemistry Chemical Physics. 2008; **10**(44):6615. DOI: 10.1039/b810189b

[23] Yang Z, Liu C, Shao C, Lin C, Liu Y. The Journal of Physical Chemistry. 2015; **119**:21852

[24] Katono K, Bessho T, Wielopolski M, Marszalek M, Moser JE, Humphry-Baker R, et al. The Journal of Physical Chemistry. 2012;**116**:16876

[25] Rostov IV, Amos RD, Kobayashi R, Scalmani G, Frisch MJ. The Journal of Physical Chemistry. 2010;**114**:5547

[26] Pedone A. Journal of Chemical Theory and Computation. 2013;**9**:4087

[27] Kurt M, Babu PC, Sundaraganesan N, Cinar M, Karabacak M. Molecular structure, vibrational, UV and NBO analysis of 4-chloro-7-nitrobenzofurazan by DFT calculations. Spectrochimica Acta Part A: Molecular and Biomolecular Spectroscopy. 2011;**79**(5):1162. DOI: 10.1016/j.saa.2011.04.037

[28] Sudha S, Sundaraganesan N, Kurt M, Cinar M, Karabacak M. FT-IR and FT-Raman spectra, vibrational assignments, NBO analysis and DFT calculations of 2-amino-4-chloro-benzonitrile. Journal of Molecular Structure. 2011;**985**(2–3):148. DOI: 10.1016/j.molstruc.2010.10.035

[29] Gutsev GL, Bauschlicher CW. Electron affinities, ionization energies, and fragmentation energies of Fe n clusters (n= 2-6): a density functional theory study. The Journal of Physical Chemistry A. 2003;**107**(36):7013. DOI: 10.1021/jp030288p

[30] Nassar MY, El-Shahat M, Khalile S, El-Desawy M, Mohamed EA. Structure investigation of mesalazine drug using thermal analyses, mass spectrometry, DFT calculations, and NBO analysis. Journal of Thermal Analysis and Calorimetry. 2014;**117**(1):463. DOI: 10.1007/s10973-014-3638-1

[31] Frisch MJ, Trucks GW, Schlegel HB, Scuseria GE, Robb MA, Fox DJ. Gaussian, Inc., 2009

[32] Kurt M, Sertbakan T, Özduran M. An experimental and theoretical study of molecular structure and vibrational

spectra of 3-and 4-pyridineboronic acid molecules by density functional theory calculations. Spectrochimica Acta Part A: Molecular and Biomolecular Spectroscopy. 2008;**70**(3):664

[33] Kistiakowsky G, Van Artsdalen E. Bromination of hydrocarbons. I. Photochemical and thermal bromination of methane and methyl bromine. Carbon-hydrogen bond strength in methane. Journal of Chemical Physics. 1944;**12**(12):469. DOI: 10.1063/1.1723896

[34] Davidson C, De Gee A, Feilzer A. The competition between the composite-dentin bond strength and the polymerization contraction stress. Journal of Dental Research. 1984;**63**(12): 1396. DOI: 10.1177/00220345840630121101

[35] Thostenson ET, Chou TW. Aligned multi-walled carbon nanotube-reinforced composites: Processing and mechanical characterization. Journal of Physics D: Applied Physics. 2002;**35**(16): L77-L80

[36] Phung Hai TA, Sugimoto R. Conjugated carbazole-thiophene copolymer: Synthesis, characterization and applications. Synthetic Metals. 2016; **220**:59

[37] Bourass M et al. The optoelectronic properties of new dyes based on thienopyrazine. Comptes Rendus Chimie. 2017;**20**(5):461

[38] Bourass M et al. DFT/TD-DFT characterization of conjugational electronic structures and spectral properties of materials based on thieno [3,2-b] [1]benzothiophene for organic photovoltaic and solar cell applications. Journal of Saudi Chemical Society. 2017; **21**(5):563. DOI: 10.1016/j.jscs.2017.01.001

[39] Gangadharan RP, Sampath KS. Natural bond orbital (NBO) population analysis of 1-azanapthalene-8-ol. Acta Physica Polonica A. 2014;**125**(1):18. DOI: 10.12693/APhysPolA.125.18

[40] Nobel NK, Bamba K, Patrice OW, Ziao N. NBO population analysis and electronic calculation of four azopyridine ruthenium complexes by DFT method. Computational Chemistry. 2017;**05**(01):51. DOI: 10.4236/cc.2017.51005

Chapter 5

Germanium on Silicon Avalanche Photodiode for High-Speed Fiber Communication

Mengyuan Huang, Kelly Magruder, Yann Malinge, Parastou Fakhimi, Hao-Hsiang Liao, David Kohen, Gregory Lovell, Wei Qian, Kiyoung Lee, Carsten Brandt, Mahtab Hakami, Yen-jung Chen, Erin Carabajal, Erle Guillermo, Seth Slavin and Ansheng Liu

Abstract

Silicon photonics is one of the promising technologies for high-speed optical fiber communications. Among various silicon photonic devices, germanium on silicon avalanche photodiode (Ge/Si APDs) received tremendous attentions because of its superior performance and integration compatibility. In 2016, normal incidence Ge/Si APD demonstrated a NRZ 10^{-12} sensitivity of −23.5 dBm at 25 Gb/s; more recently, a waveguide-integrated Ge/Si APD receiver presents a 106Gb/s PAM4 sensitivity of −18.9 dBm. These results are best reported performance among all APD-based devices, and these breakthroughs are mainly benefited from Ge/Si APD's structure and material characteristics. Ge/Si APD adopts a separated charge-absorption-multiplication (SCAM) structure with a pure Ge absorber and an intrinsic Si avalanche layer. Since, Si is one of well-known best avalanche materials with large gain-bandwidth products and low ionization noise ratio, which make Ge/Si APDs demonstrating superior performance at high data rates. Moreover, this Si-based device is manufactured by standard CMOS foundries and is process-compatible with other silicon photonic devices including silicon-based waveguides, demux, hybrid, etc. This advantage simplifies the assembly of photonic systems and makes a large-scale integrated silicon photonic chip possible, which provides compact solutions for high-density communication systems. In this chapter, we review recent progresses on Ge/Si APD structure design, material, and performance.

Keywords: silicon photonics, avalanche photodiode, high-speed, germanium on silicon, Fiber communication

1. Introduction

Data traffic grows exponentially in last few years [1–4]. From 2018 to now, the global digital data creation experiences a compound annual growth rate of 25% [5], and the estimated annual data traffic will be four times larger at 2025 [6]. All these fast-expending data demands are accommodated by fiber communication systems. In nowadays, more than 20% interconnects in data centers already reach operating data rate of 100 Gb/s per lane [7]. To maximize bandwidth efficiency, latest systems adopt various technologies including wavelength division-multiplexing (WDM), complex modulations (PAM4), etc., which are inevitable to bring more penalties from original ones. For instance, CWDM technology typically causes ~4.5 dB loss on optical powers [8] and complex PAM4 modulation brings 4.8 dB loss compared to simple NRZ modulation [9, 10]. Therefore, a high-speed and high-sensitivity solution is essential to compensate these extra losses and to support the upgrades of data and telecommunication systems to new generation.

Avalanche photodiode (APD) and coherent detection are most promising solutions with gains. Compared to coherent detection, APD receivers have a variety of advantages including smaller size, lower power consumptions, better latency, and lower cost [11]. These merits make the large-scale deployment of APDs into high-sensitivity systems. One example is the 29 dB-link budget requirement in 10 Gb/s passive optical networks (PON) [12]. Photodiodes (PD) cannot meet such high requirement, and APDs provide extra ~8 dB gains at data rate of 10 Gb/s or 25 Gb/s [13–15]. Thus, millions of APD devices are utilized in PON systems for fiber-to-home applications. However, when bandwidth is increased to 100 Gb/s in today, traditional APD's additional gain drops to 2–4 dB [16, 17] that is related to InP-based material fundamental limitations (poor gain-bandwidth products) [18]. On the other hand, Ge/Si APD demonstrated a great linear gain-bandwidth products of 340 GHz [19], which are the fundamental reasons of Ge/Si APD presenting better performance at higher data rates such as 100 Gb/s and beyond.

2. Ge/Si APD structure and responsivity improvement

The first demonstration of high-speed Ge/Si APD was completed by Intel in 2007 [20], and that device shows a 3 dB bandwidth of 7 GHz at gain = 1. This device is a top-illuminated device on bulk Si wafer, and device's main functional layers include a heavily n-doped silicon contact layer, a 0.5 μm thick intrinsic silicon multiplication layer, a 0.1 μm thick p-type silicon charge layer, a 1 μm thick pure germanium absorption layer, and a 0.1 μm heavily p-doped germanium contact layer as shown in **Figure 1**.

This APD reports a responsivity of 0.52 A/W at 1310 nm, which is obvious worse than 10 Gb/s III-V APD's responsivity of 0.85 A/W [21]. The fundamental reason is that Ge's absorption coefficients are worse than $In_{0.53}Ga_{0.47}As$ at O-band wavelengths. To improve Ge/Si device's responsivity, there have two promising designs including resonant cavity-enhanced (RCE) structure and waveguide-integrated structure.

2.1 RCE normal incidence Ge/Si avalanche photodiode

RCE photodiode structure is studied by several groups using various materials [22, 23]. The principle is adding both top and bottom reflectors to photodiodes' active layers for forming a Fabry-Perot resonant cavity. The RCE design provides a multi-pass absorption scheme for responsivity improvement but no impact on high-speed characteristics. RCE photodiode's quantum efficiency is given by following formula [23]:

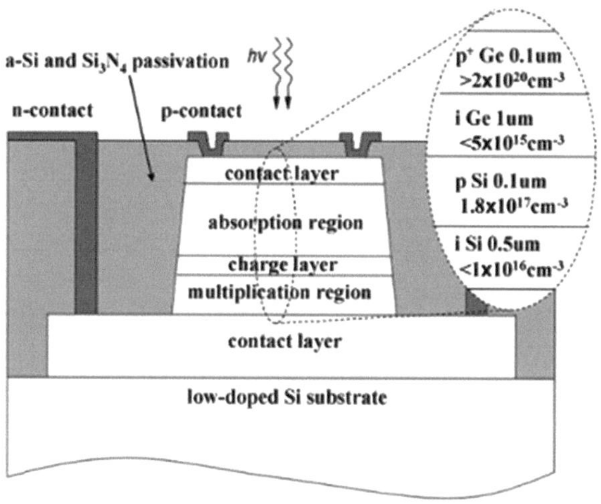

Figure 1.
Cross section of Ge/Si APD on bulk Si substrate [20].

$$\eta_{RCE} = \left(\frac{1 + R_2 \exp(-\alpha d)}{1 - 2\sqrt{R_1 R_2} \exp(-\alpha d) \cos(2\beta L) + R_1 R_2 \exp(-\alpha d)}\right)(1 - R_1)(1 - \exp(-\alpha d)) \tag{1}$$

Here, R_1 is the reflectivity at top surface of RCE cavity, and R_2 is the reflectivity at bottom surface, α is absorption coefficient, d is absorber thickness, L is the cavity length, and β is the propagation constant.

Considering photodiode's top surface is typically made by antireflection (AR) coating with low reflectivity <5%, a high reflectivity surface at bottom is critical for enhancing responsivity. Several papers reported Ge/Si RCE devices by using different bottom reflectors [24–27]. One common solution is using SOI substrate: a reflection happens at the interface between Si and buried oxide (BOX) layers because of large refractive index gap. Especially for double SOI substrate, the bottom reflectivity can reach >90% with optimized silicon and oxide thicknesses [24]. Alternative method is to use CMOS-compatible metals such as aluminum with >95% reflectivity at optical communication wavelengths (**Figure 2**) [28].

Figure 3 presented measured 25 Gb/s NIAPDs' responsivity with and without RCE structure. We can clearly see that: after achieving a > 100% improvement on responsivity at peak wavelengths (1310–1314 nm), the RCE device's optical bandwidth becomes narrower like full width at half maximum ~40 nm.

Resonated photodiode's optical bandwidth is related by free spectral range (*FSR*) and Finesse (*F*), the full width at half maximum (FWHM) is given by following equation [29]:

$$FWHM = \frac{FSR}{F} = \frac{\lambda^2}{2n_{eff}L} \times \frac{1 - \sqrt{R_1 R_2} e^{-\alpha d}}{\pi (R_1 R_2)^{1/4} e^{-\frac{\alpha d}{2}}} \tag{2}$$

Here, λ is operating wavelength, n_{eff} is the effective refractive index, and L is the total length of RCE cavity. Since, Ge/Si photodetector has thin absorption layer (e.g.

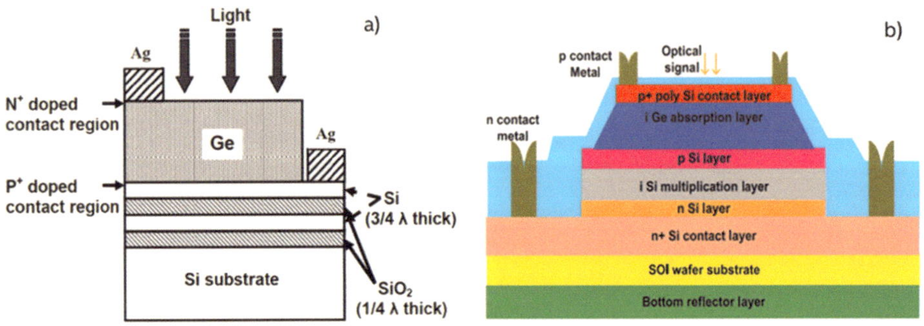

Figure 2.
a) Ge/Si PD on DSOI [24]; b) Ge/Si APD with metal reflector [27].

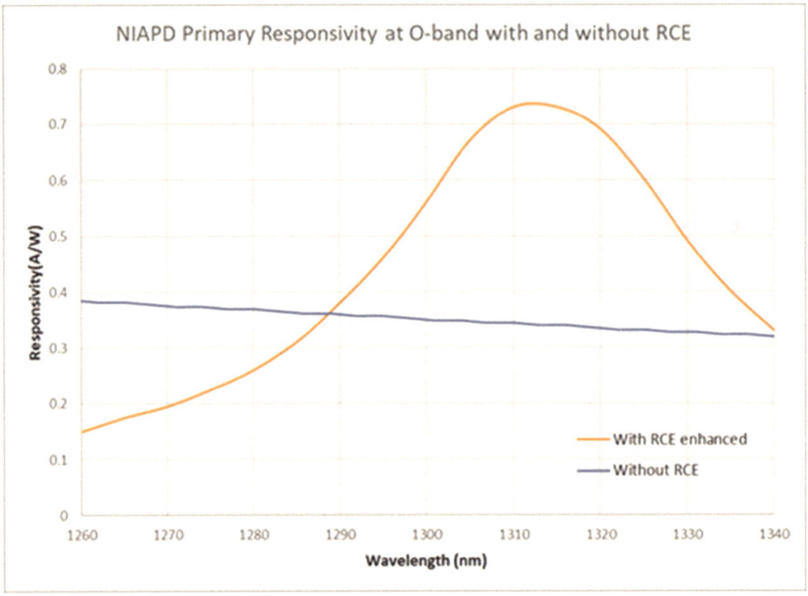

Figure 3.
NIAPD responsivity with and without RCE at CWDM4 wavelengths.

<600 nm) for high-speed operations, which brings a narrow optical 3 dB bandwidth (e.g. <50 nm) and not suitable for current 2–10 km data centers systems using multiples wavelengths like CWDM4 [30, 31].

2.2 Waveguide-integrated Ge/Si avalanche photodiode

Waveguide structure has better solutions with high bandwidth and flat optical spectrum simultaneously. The fundamental improvement of waveguide photodiode is that photon-generated carriers move into a different direction from optical propagations, which breaks the trade-off between 3 dB bandwidth and responsivity on normal incident devices.

Figure 4 presents evanescent coupling Ge/Si WGAPD devices [32]. In this design, on-chip optical power is confined by Si waveguide and propagates to WGAPD. Ge/Si WGAPD's critical layers—Si multiplication, charge, and Ge absorption layers—are grown on these Si waveguide layers. Therefore, it has an unavoidable height between silicon waveguide and Ge absorber, which degrades coupling efficiency and responsivity [33]. For instance, evanescent Ge/Si WGAPD only shows 0.6 A/W responsivity at 1550 nm even with 50 μm length Ge absorber [34].

Optimized waveguide design is recently reported recess-type structure [2], which solves height difference by etching a recess into Si waveguide and selectively depositing Si and Ge films into the recess. Therefore, the gap between Si waveguide and Ge absorber is minimized; as a result, evanescent device's responsivity is improved >0.75 A/W [2]. **Figure 5** presents the schematic cross section of recess-type Ge/Si waveguide APD.

Another advantage of this recess structure is using thick top Si waveguide, which supports both TE and TM modes with low propagation losses. This structure only needs one-side input optical waveguide without polarization rotator-splitter, which reduces optical losses and improves external responsivity and overall performance.

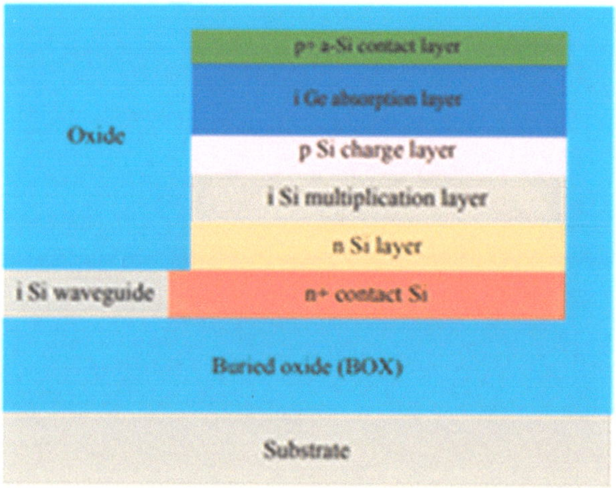

Figure 4.
Waveguide-integrated Ge/Si APD using evanescent coupling [32].

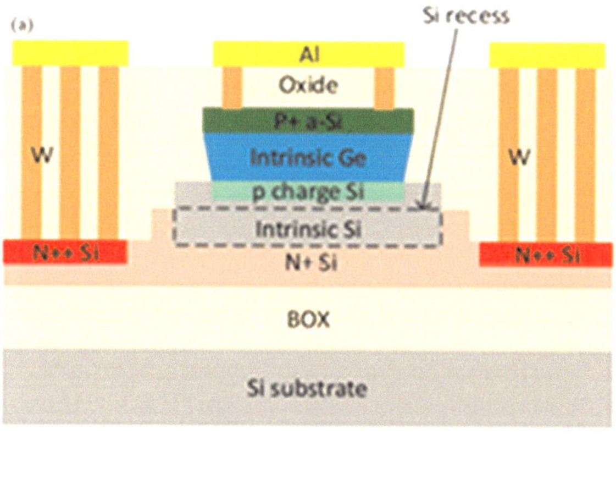

Figure 5.
Recess-type waveguide-integrated Ge/Si avalanche photodiode [2].

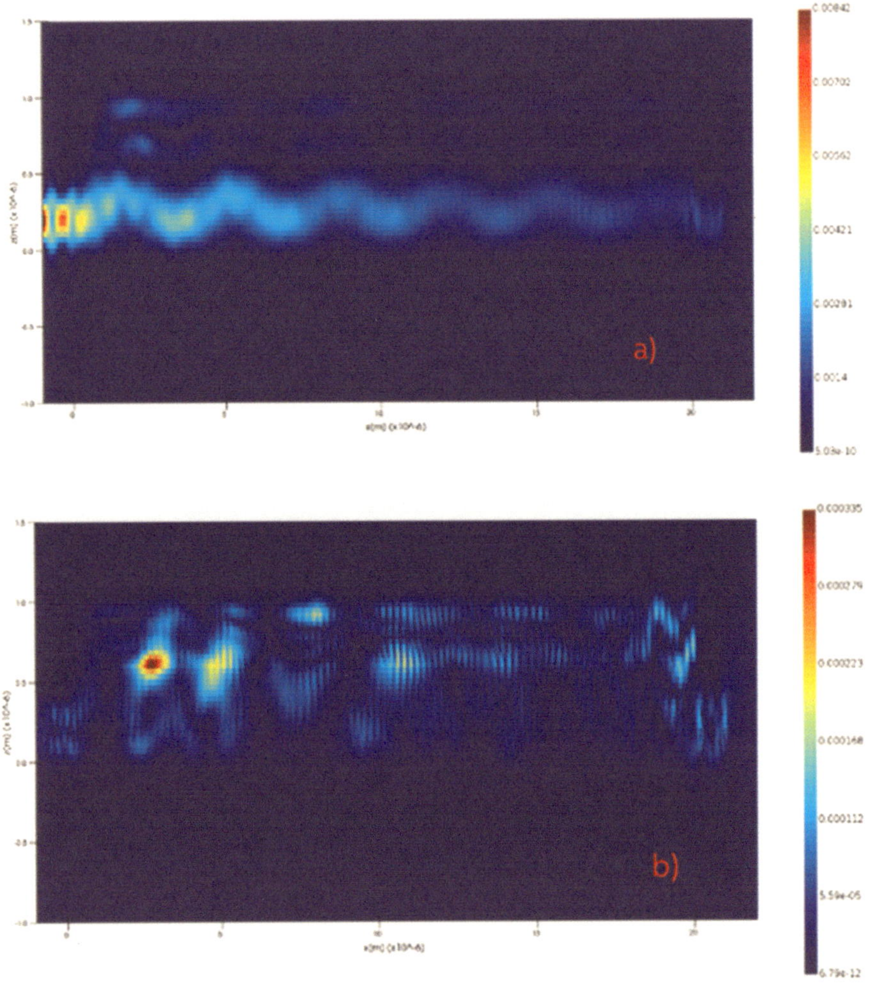

Figure 6.
a) TE mode and b) TM mode propagation in recess-type Ge/Si WGAPD.

Figure 6 shows technology computer-aided design (TCAD) simulations of both TE and TM modes propagations in this recess-type WGAPD structure:

Figure 7 presents measured primary responsivity (gain = 1) of recess-type Ge/Si APD. This result is extracted by WGPD structure using similar film thickness as WGAPD, which shows a flat responsivity from 1260 to 1340 nm wavelengths and suitable for CWDM4 system applications.

3. Ge/Si interface and Ge/Si APD RF performance

Ge/Si devices' RF performance is discussed by several groups [35, 36]. However, several critical effects were not comprehensively studied. For instance, Ge/Si interface characteristics play an important role on APDs' RF performance, because all photon generated electrons must across this interface for electrical amplifications. Because of 4.2% lattice mismatch between Ge and Si substrate, a variety of dislocations are

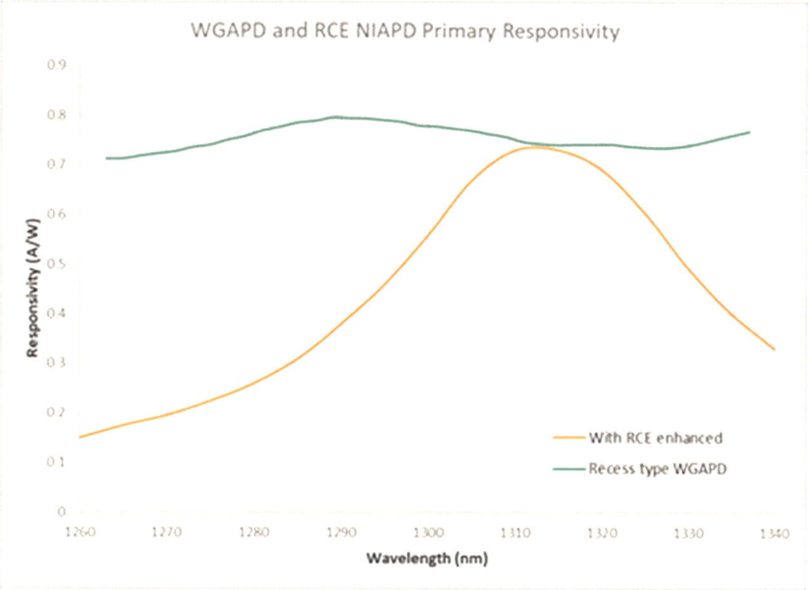

Figure 7.
WGAPD and NIAPD primary responsivity at CWDM4 wavelengths.

formed inside Ge epitaxial layer. Especially at Ge/Si interface, threading dislocation density typically reaches $1\times10^{10}/cm^2$ [37]; in additional, volumes of misfit dislocations [38] and lots of unpaired dangling bonds [39] locate at the interface region. These characteristics cause several undesirable impacts on APDs' RF performance.

3.1 Shallow-level defect states and its impact on RF performance

For Ge on Si films, different groups reported the observations of dislocations-related deep-level defect states with energy level of 0.37 eV [40, 41]. This defect state captures free electrons for non-radiative recombination and contributes to devices' dark currents [42].

Different from deep-level defect states, shallow defect level state also found inside Ge on Si film, which has an energy level of 0.02 eV above valence band and behaviors as acceptors [43]. The fundamental formation of this shallow-level defect state could be explained by unpaired dangling bonds at Ge/Si interface. Since Si lattice constant (5.431 Å) is smaller than Ge (5.658 Å), many Si atoms ($\sim10^{13}$ atoms/cm^2) are unpaired at Ge/Si interface, which attract electrons like p-type dopants for forming a stable covalence bond (**Figure 8**).

Several groups reported the existence of this shallow-level defect states by using various Ge growth methods including ultra-high vacuum chemical vapor deposition (UHV-CVD), reduced pressure chemical vapor deposition (RPCVD), and sputter [43–45]. **Figure 9** provides a summary of shallow-level defect state's effective concentration. Based on reported data, defect effective carrier concentrations are increasing near Ge/Si interface region. This phenomenon has similar trend of material data: there have more unpaired bonds and more dislocations close to Ge/Si interface.

This shallow-level defect state brings serious impacts on device's electrical performance. First, high concentration reduces the mobility of photon-generated carriers. For example, although hole mobility reaches 1900 cm^{2}V^{-1} in intrinsic Ge film [46],

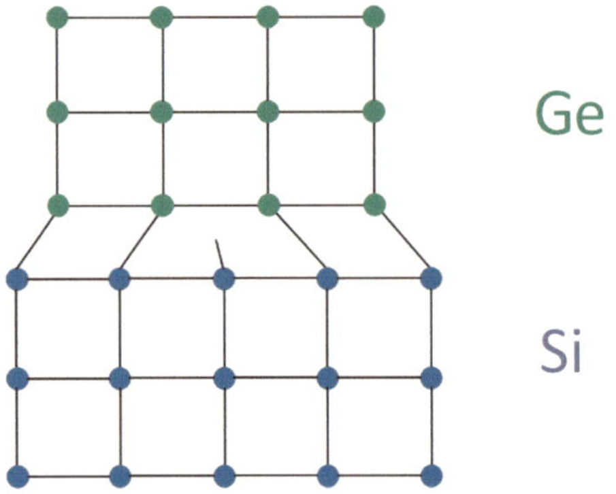

Figure 8.
Ge/Si interface atom distribution and dangling bonds.

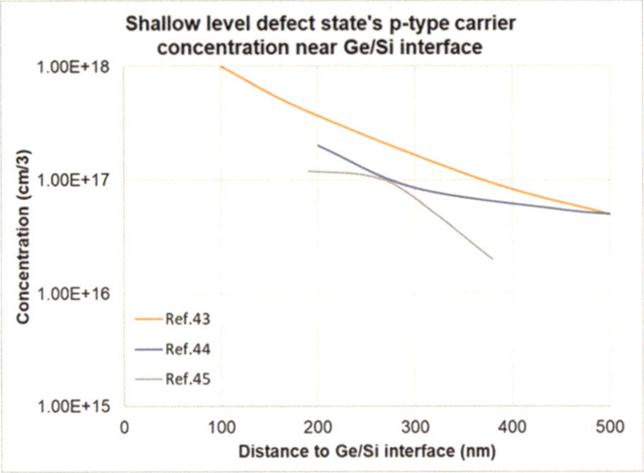

Figure 9.
Shallow-level defect states' concentration at Ge/Si interface ref. [43–45].

the number drops to only ~45 cm^2V^{-1} [43] at defect carrier concentration of 2×10^{17} cm^{-3}. Such mobility degradation significantly increases diffusion time in Ge and then reduces 3 dB bandwidth.

Based on reported data, a 250–350 nm thick Ge with concentration > $10^{17}/cm^3$ is found close to Ge/S interface. This p-type doped region requires high bias for depletion and for carrier reaching saturation velocity. **Figure 10a)** and **b)** present the E-field distribution comparison with and without shallow-level defect state:

Figure 10c) presents local electrical field intensity dependence on defect states' carrier concentrations. Given a 250 nm-thick Ge layer with 1×10^{17} cm^{-3} concentration, the local E-field reaches ~150 kV/cm which can easily charge carriers energy beyond avalanche threshold inside Ge [47, 48]. Considering poor Ge's ionization coefficient ratio, e. g. ~1 [49], APD device's transit time bandwidth is significantly degraded with avalanching in Ge layer [50]. If this defect effective concentration is even larger like >2×10^{17} cm^{-3}, the depleted region will have an E-field >200 kV/cm with even worse impacts like huge leakage currents from band-to-band tunneling effects [51]. These problems completely ruin high-speed APD devices. Therefore, it is extremely critical to control of Ge/Si interface quality and many methods—like dilute HF-last pre-clean, high-temp pre-bake, low-temp pre-bake with HCl clean, in situ process—had been studied [52–55].

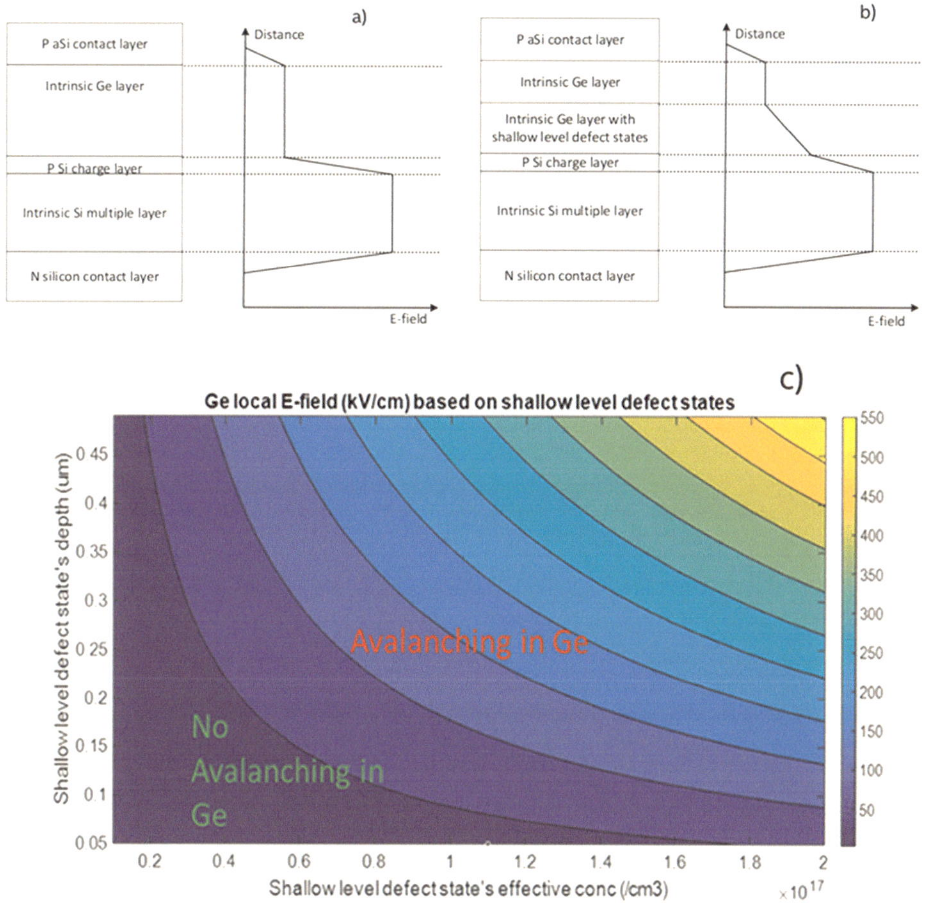

Figure 10.
a) Ge/Si APD's ideal E-field; b) Ge/Si APD's E-field with shallow-level defect states at Ge/Si interface; c) E-field need to deplete shallow-level defect states.

3.2 Ge/Si heterojunction and impact on RF performance

Ge/Si heterojunction is important for understanding carrier transport mechanisms inside APD devices. Because Ge has an electron affinity of 4.0 eV close to Si value of 4.05 eV, different types of Ge/Si band alignment had been reported. Several studies show Ge/Si type-I heterojunction [56–58], and other papers reported type-II heterojunction [59–61]. To better understand this issue, we investigated free carrier distributions at Ge/Si interface by using two different silicon substrates (either heavily p-doped or heavily n-doped) (**Figure 11**).

Based on measured data, samples on either N-substrate or P-substrate show carriers' accumulations at Ge/Si interface, which proves type-I heterojunctions because type-II band offset has no barrier for electrons' accumulations.

Device-level data also support type-I heterojunction. **Figure 12** shows measured WGAPD S21 curves under different biases. At operating with bias lower than breakdown voltage (V_{br}), APD's S21 response is decreasing at high frequency with a 3 dB bandwidth of 27.5 GHz. While, at high bias beyond breakdown voltage, RF enhancement and peaking happen on APD's S21 curves and then 3 dB bandwidths are

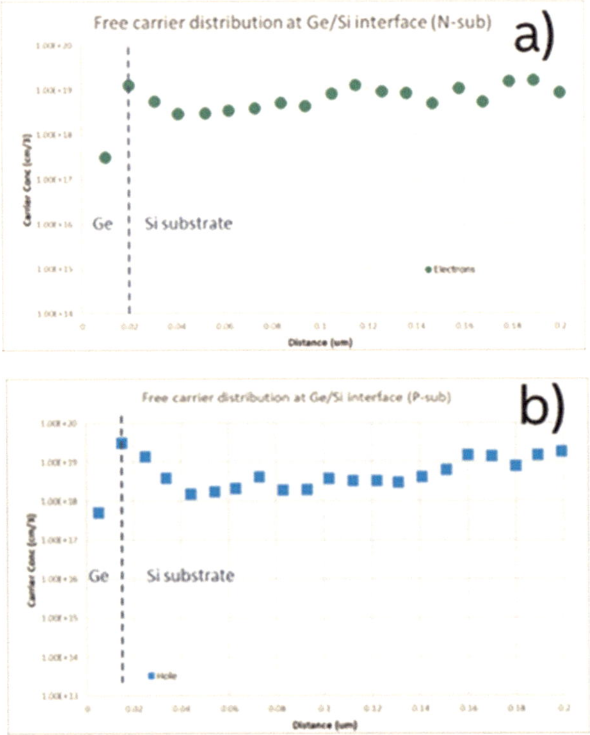

Figure 11.
Measured free carrier distribution at Ge/Si interface: a) N-sub and b) P-sub.

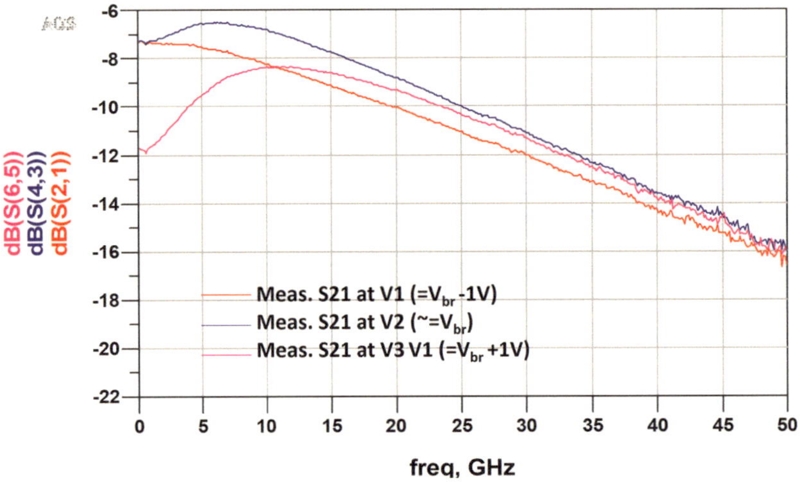

Figure 12.
Measured S21 curves of Ge/Si APD under different bias.

increased significantly. These frequency response peaking and bandwidth enhancements are related to negative differential resistance (NDR) effect [62, 63].

NDR effect has been frequently observed in quantum wells photonic devices, and tunneling effect is used to explain this phenomenon [64, 65]. However, Ge/Si APD utilizes bulk Ge on Si without quantum well; thus, the only possible energy barrier is the conduction band offset from type-I Ge/Si heterojunction. Another evidence is that NDR enhanced 3 dB bandwidths only happened at Ge/Si APD with bias ranges beyond V_{br} causing a local high E-field, which is significantly increasing the possibility of electrons' tunneling at Ge/Si interface as following diagram (**Figure 13**):

3.3 Re-visit Ge/Si APD's gain-bandwidth product

Because of the NDR effect existence, it is necessary to revisit Ge/Si APDs' reported gain-bandwidth products. **Table 1** summarized published Ge/Si APD gain-bandwidth data with and without NDR effect.

Based on reported data, the same APD device's shows obviously larger gain-bandwidth product under NDR effect. However, such high-performance region is not suitable for high-speed communication applications. One reason is that APDs' S21 responses have frequencies-related peaking and drops, which bring drawbacks including variations in group delay [72] and receiver's DC responsivity losses. Another reason is that NDR effect only happens at bias $>V_{br}$, which led to high dark currents (e.g. >1 mA) and high shot noise. Moreover, mA-level dark currents cannot meet several commercial transceivers' requirements such as loss of signals (LOS) under weak input optical powers [73]. In summary, APD under NDR is not suitable for high-speed communication systems, but it can be applied to single-photon detection like Geiger-mode operations [74].

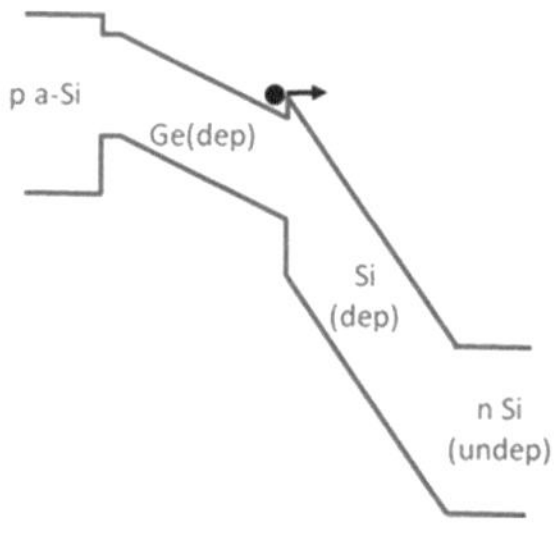

Figure 13.
Ge/Si APD band structure under high E-field and NDR effect.

Operating type	Gain-bandwidth product (without NDR)	Gain-bandwidth product (with NDR)	Ref
Ge/Si NIAPD	340 GHz	—	[16]
	—	840 GHz	[66]
	258 GHz	—	[67]
	300 GHz	450 GHz	[68]
Ge/Si WGAPD	360 GHz	—	[69]
	—	310 GHz	[70]
	280 GHz	410 GHz	[71]
	300 GHz	—	[2]
Applications	High-speed communication	Single photo detection	

Table 1.
Reported Ge/Si APD w/ and w/o NDR effect.

3.4 Ge/Si APD equivalent circuit model

High-speed photodiodes' or APD's equivalent circuit model had been investigated by different groups [70, 71, 75]. Typically, the circuit model includes transit time and RC-time circuit models. The RC-time model in equivalent circuit contains both APD junction parameters and parasitic parameters, which can be extracted by using fitting results of S22 as shown in **Figure 14(b)**. Different from III-V APDs on semi-insulating substrate, Ge/Si APD device is developed on a SOI wafer with a thick buried oxide layer (1-3 μm) with a great RF isolation. Thus, parasitic components have weak impact on Ge/Si APD's RF characteristics, and **Table 2** presents values extracted by circuit model.

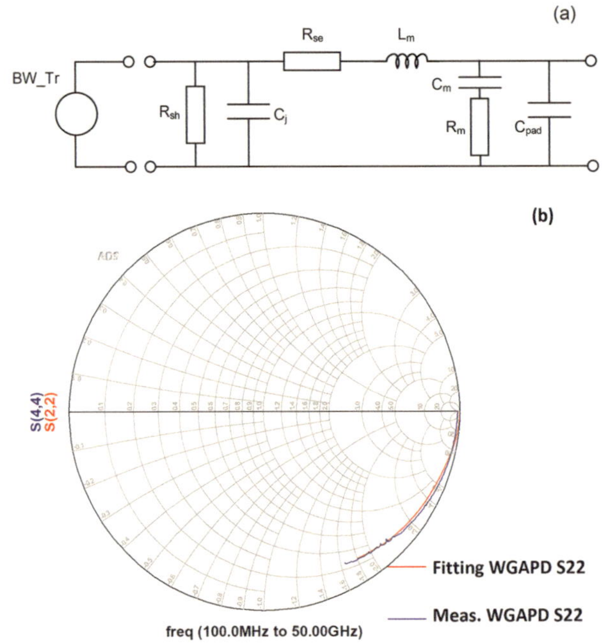

Figure 14.
a) Ge/Si APD circuit model; b) Ge/Si APD S22 measured and fitting results.

Parameter	Symbol	Value
Shunt resistance	R_{sh}	4.6 MΩ
Junction capacitance	C_j	22.4 fF
Series resistance	R_{se}	13.4 Ω
Metal inductance	L_m	140 pH
Pad capacitance	C_{pad}	5.5 fF
Metal/sub-capacitance	C_m	6.4 fF
Metal/sub-resistance	R_m	1.6 kΩ

Table 2.
106 Gb/s Ge/Si WGAPD junction and parasitic parameters.

4. Ge/Si APD receiver performance and applications

4.1 Ge/Si APD receiver's RF model

The RF model of entire Ge/Si APD receiver is the combination of APD equivalent circuit model, transimpedance amplifier (TIA) model, interconnect model, and evaluation board's model shown as **Figure 15(a)**. The interconnect between APD and TIA is generally achieved by a short gold wire, which typically can be considered a small inductor [76]. With proper modeling of receiver S-parameters, the simulation S21 curves show great matching of measured results as **Figure 15(b)**.

4.2 Ge/Si APD receiver's sensitivity and overload performance

Sensitivity and overload performance are two mostly critical system-level specifications for high-speed receivers. For sensitivity, it is typically defined at certain signal-to-noise ratio (SNR). APD receiver's signal-to-noise ratio (SNR) is related to several parameters including APD responsivity, gain and dark current, TIA noise current, and receiver bandwidth like following equation [77]:

$$\frac{S}{N} = \frac{\left(\frac{1}{2}\right)\left(\frac{q\eta P_S}{h\nu}\right)^2}{2q\left(I_{dark} + I_{photo}\right)FB + \frac{<i_{TIA}^2>}{M^2}} \tag{3}$$

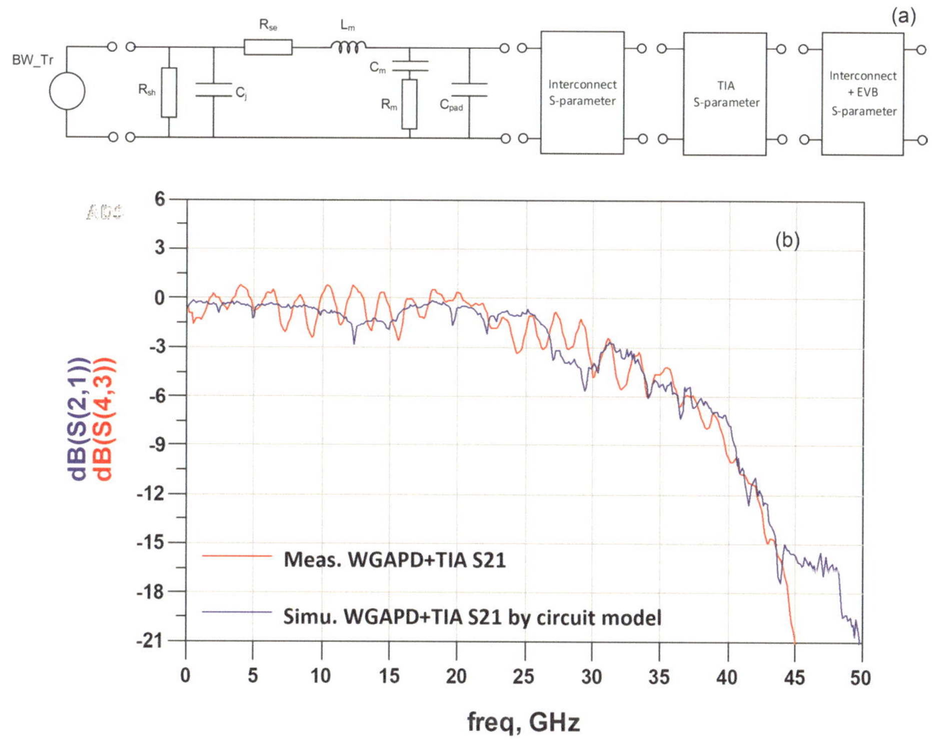

Figure 15.
a) Ge/Si APD receiver model; b) Ge/Si APD receiver measured S21 and fitting results.

Here, F is excess noise factor and related to APD gain (M) and ionization coefficient ratio k:

$$F = kM + \left(2 - \frac{1}{M}\right) \times (1 - k) \tag{4}$$

From Eqs. (3) and (4), TIA noise current is inversely proportional to APD gains, and excess noise factor is increasing with gain. APD SNR presents a parabolic curve with gain increasing as shown in **Figure 16**:

On the other hand, the receiver overload happens at rms photocurrent value larger than the threshold of TIA [79]. And rms photo current is given by [80]:

$$i_p = \frac{q\eta m P_o M}{\sqrt{2}h\nu} \tag{5}$$

For APD operating gain range of 1–8, APD receiver's sensitivity P_s is improved at larger gain, but overload P_o is degraded. Therefore, a trade-off is necessary for balancing sensitivity and overload. **Figure 17** presents our recess-type Ge/Si APD BER results. Our receiver unstressed sensitivity reaches −18.9 dBm at BER 2×10^{-4}, which provide >5 dB margin compared to 100 Gb/s per lane 40 km sensitivity specification (−13.8 dBm) [81]. Moreover, our receiver presents a flat BER below 10^{-7} with input optical power up to 0 dBm that provide enough margins for overload specifications (−2.6 dBm) [81].

As shown in **Figure 17**, Ge/Si WGAPD receiver provides ~8 dB margin compared to PIN photodiode solution. It provides record ~19 dB dynamic range for 106 Gb/s operating, which supports both short reach applications like back-to-back and 40 km long-reach applications.

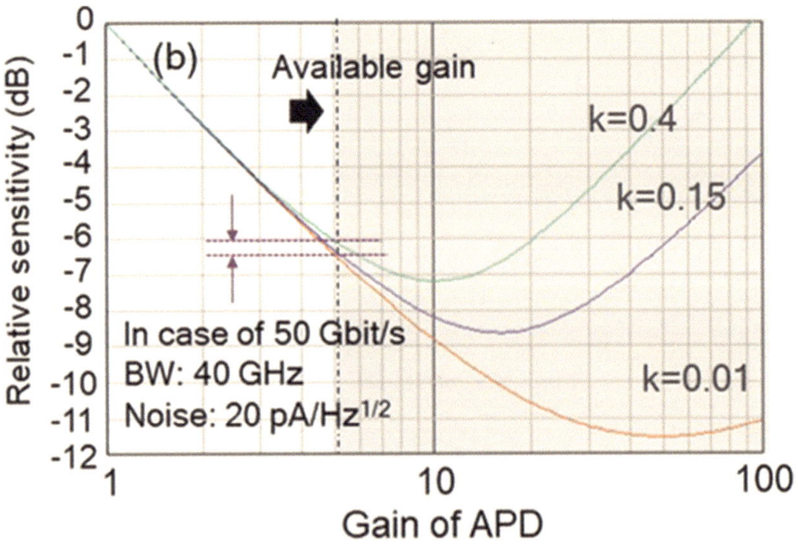

Figure 16.
50 Gb/s APD receiver sensitivity vs. gain [78].

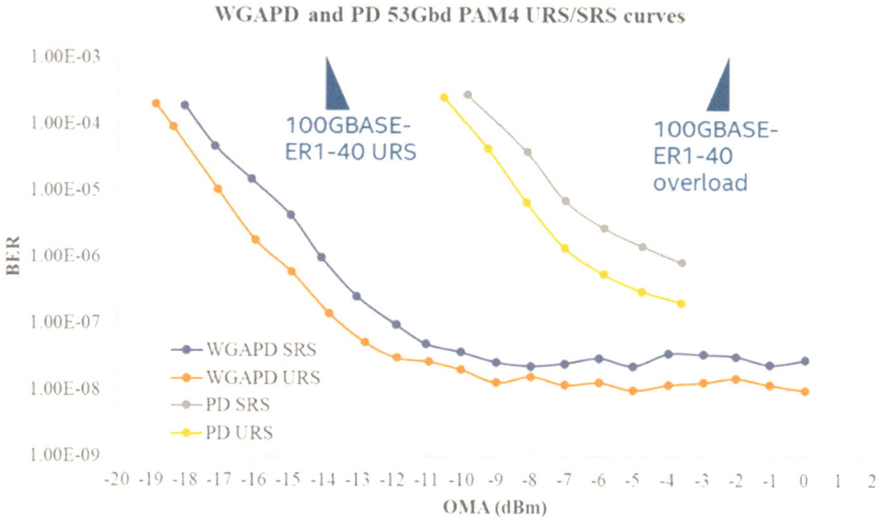

Figure 17.
106 Gb/s WGAPD and PD receiver performance vs. IEEE 100G-ER1-40 specification [2].

4.3 Ge/Si APD receiver's current and future commercial applications

5G mobile network infrastructures are deployed worldwide. The data rate reaches 50 Gb/s at 5G front haul [82] and increases to 100 Gb/s at 5G middle/back hauls [83]. These applications require high-sensitivity devices for achieving passive networks without optical amplifiers. Because of Ge/Si superior sensitivity and overload at 100 Gb/s, millions of Ge/Si APD receivers are deployed in 5G wireless systems in last few years [84]. Moreover, Ge/Si APDs have more important applications including next-generation fiber communication systems. One example is coming 800G/1.6 T interconnects inside data centers with 200 Gb/s per lane data rate. Such high-bandwidth

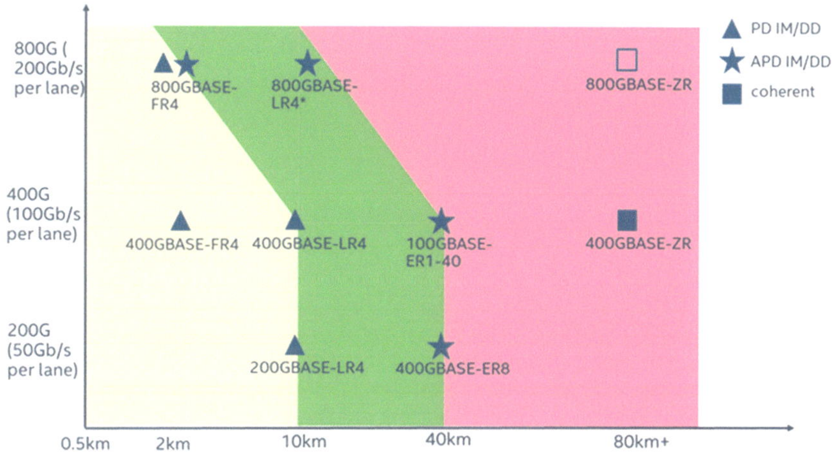

Figure 18.
IEEE/MSA standards and receiver solutions (current and future).

applications degrade receiver's sensitivities, and photodiodes can only cover operation distances within 2 km [85, 86]. For other 2 km and beyond applications, APD receivers become one of most attractive solutions because of their high performance. **Figure 18** presents various IEEE standards and related receivers' solutions with different data rates and operation distances:

5. Conclusion

In this chapter, we study structure evolutions from Ge/Si NIAPD to recess-type Ge/Si WGAPD. The latest structure provides high responsivity and flat operation spectrum suitable for WDM schemes. Moreover, we comprehensively analyze Ge/Si APD's essential characteristics including shallow-level defect states, heterojunction type, NDR effect, and gain-bandwidth products. These studies provide fundamental understandings of Ge on Si materials and related APD's performance. Furthermore, we review our Ge/Si APD receivers' sensitivity and overload including receivers' applications in current and future optical fiber and data communications.

Author details

Mengyuan Huang[1*], Kelly Magruder[2], Yann Malinge[1], Parastou Fakhimi[2], Hao-Hsiang Liao[1], David Kohen[2], Gregory Lovell[1], Wei Qian[1], Kiyoung Lee[2], Carsten Brandt[2], Mahtab Hakami[1], Yen-jung Chen[1], Erin Carabajal[2], Erle Guillermo[2], Seth Slavin[2] and Ansheng Liu[1]

1 Intel Corporation, Santa Clara, CA, USA

2 Intel Corporation, Rio Rancho, NM, USA

*Address all correspondence to: mengyuan.huang@intel.com

References

[1] Huang M et al. Breakthrough of 25Gb/s germanium on silicon avalanche photodiode. In: Optical Fiber Communication Conference, Paper Tu2D.2. 2016

[2] Huang M et al. Recess-type waveguide integrated germanium on silicon avalanche photodiode. In: Optical Fiber Communication Conference, Paper F2C.3. 2021

[3] Campbell JC et al. Recent advances in avalanche photodiodes. In: Optical Fiber Communications Conference and Exhibition (OFC), M3C.1. 2015

[4] Zhang J et al. 64Gb/s PAM4 and 160Gb/s 16QAM modulation reception using a low-voltage Si-Ge waveguide-integrated APD. Optics Express. 2020; **28**(16):23266-23273

[5] Reinsel D et al. The Digitization of the World from Edge to Core, An IDC White Paper. 2018

[6] LightCounting Ethernet Transceivers Forecast. LightCounting, March 2020

[7] Yole: Optical transceiver for Datacom & Telecom, Marketing and Technology Report. 2022

[8] Mobile Optical Pluggables (MOPA), Technical Paper. 2022

[9] Szczerba K et al. 4-PAM for high-speed short-range optical communications. Journal of Optical Communications and Networking. 2012; **4**(11):885-894

[10] Cole C. Ideal SNR penalties, IEEE 802.3 interim meeting, 400Gb/s Ethernet task force. 2014

[11] Yoshimatsu T et al. Compact and high-sensitivity 100-Gb/s (4 × 25 Gb/s) APD-ROSA with a LAN-WDM PLC demultiplexer. Optics Express. 2012;**20** (26):B393-B398

[12] Harstead E et al. From 25 Gb/s to 50 Gb/s TDM PON: Transceiver architectures, their performance, standardization aspects, and cost modeling. Journal of Optical Communications and Networking. 2020; **12**(9):D17-D26

[13] Nakagawa J et al. First Demonstration of 10G-EPON and GE-PON Co-Existing System Employing Dual-Rate Burst-Mode 3R Transceiver, Optical Fiber Communication Conference, paper PDPD10, 2010

[14] Nada M et al. High-sensitivity 25 Gbit/s avalanche photodiode receiver optical sub-assembly for 40 km transmission. Electronics Letters. 2012; **48**(13):777–778

[15] Shiba K et al. Theoretical and experimental study on waveguide avalanche photodiodes with an Undepleted absorption layer for 25-Gb/s operation. Journal of Lightwave Technology. 2011;**29**(2):153–161

[16] Nada M et al. A 42-GHz bandwidth avalanche photodiodes based on III-V compounds for 106-Gbit/s PAM4 applications. Journal of Lightwave Technology. 2019;**37**(2):260-265

[17] Nada M et al. 106-Gbit/s PAM4 40-Km Transmission Using an Avalanche Photodiode with 42-GHz Bandwidth, Optical Fiber Communication Conference, Paper W4D.2. 2018

[18] Campbell JC et al. Recent advances in avalanche photodiodes. IEEE Journal of Selected Topics in Quantum Electronics. 2004;**10**(4):777–787

[19] Kang Y et al. Monolithic germanium/ silicon avalanche photodiodes with 340 GHz gain–bandwidth product. Nature Photonics. 2009;**3**:59–63

[20] Kang Y et al. Ge/Si Avalanche Photodiodes for 1.3μm Optical Fiber Links, IEEE International Conference on Group IV Photonics, Paper 2007 FD6. 2007

[21] Levine BF et al. −29dBm Sensitivity, InAlAs APD-Based Receiver for 10Gb/s Long-Haul (LR-2) Applications, Optical Fiber Communication Conference, Paper OFM5. 2005

[22] Kishino K et al. Resonant cavity-enhanced (RCE) photodetector. IEEE Journal of Quantum Electronics. 1991;**27** (8):2025–2034

[23] Ünlü MS et al. A theoretical study of resonant cavity-enhanced photodetectors with Ge and Si active regions. Applied Physics. 1992;**71**(8): 4049-4058

[24] Dosunmu OI et al. Germanium on double-SOI photodetectors for 1550 nm operation. In: Proc. SPIE 5353, Semiconductor Photodetectors. 2004

[25] Huang M et al. 25Gb/s normal incident Ge/Si avalanche photodiode. In: European Conference on Optical Communication (ECOC). Cannes, France: We.2.4.4. 2014

[26] Huang M et al. Development of Si photonics technology: Ge/Si avalanche photodiode for PON applications. In: Optical Fiber Communication Conference. 2014

[27] Huang M et al. Germanium on silicon avalanche photodiode. IEEE Journal of selected Topics of Quantum Electonics. 2018;**24**(2):1-11. (Article ID: 3800911)

[28] Bennett HE et al. Infrared reflectance of Aluminum evaporated in ultra-high vacuum. Journal of the Optical Society of America. 1963;**53**(9): 1089-1095

[29] Ünlü MS, Strite S. Resonant cavity enhanced photonic devices. Journal of Applied Physics. 1995;**78**:2

[30] IEEE 400GBASE-LR4 specification

[31] IEEE 400GBASE-FR4 specification

[32] Huang M et al. 56GHz Waveguide Ge/Si Avalanche Photodiode, Optical Fiber Communication Conference, Paper W4D.6. 2018

[33] Wang X, Liu J. Step-coupler for efficient waveguide coupling to Ge/Si avalanche photodetectors. IEEE Photonics Technology Letters. 2011;**23** (3):146-148

[34] Kang Y et al. High performance Ge/ Si avalanche photodiodes development in intel. In: Optical Fiber Communication Conference, Paper OWZ1. 2011

[35] Piels M, Bowers JE. 40 GHz Si/Ge uni-traveling carrier waveguide photodiode. Journal of Lightwave Technology. 2014;**32**(20):3502-3508

[36] Ishikawa Y, Wada K. Near-infrared Ge photodiodes for Si photonics: Operation frequency and an approach for the future. IEEE Photonics Journal. 2010;**2**(3):306-320

[37] Wang G et al. A model of threading dislocation density in strain-relaxed Ge and GaAs epitaxial films on Si (100). Applied Physics Letters. 2009;**94**: 102115-1–102115-3

[38] Kim M, Hashemi P, Hoyt JL. Increased critical thickness for high Ge-

content strained SiGe-on-Si using selective epitaxial growth. Applied Physics Letters. 2010;**97**(26):262106-1–262106-3

[39] Oldham WG, Milnes AG. Interface states in abrupt semiconductor heterojunctions. Solid-State Electronics. 1964;7(2):153-165

[40] Claeys C, Simoen E. Germanium-Based Technologies from Materials to Devices. Elsevier; 2007. ISBN: 978-0-08-044953-1

[41] Patel N. Understanding Defects in Germanium and Silicon for Optoelectronic Energy Conversion [MIT Ph.D thesis]. 2016

[42] Giovane L et al. Correlation between leakage current density and threading dislocation density in SiGe p-i-n diodes grown on relaxed graded buffer layers. Applied Physics Letters. 2001;**78**:541-543

[43] Yeh W et al. Sputter epitaxial growth of flat germanium film with low threading-dislocation density on silicon (001). ECS Journal of Solid State Science and Technology. 2014;**3**(10): Q195-Q199

[44] Ohmachi Y et al. The heteroepitaxy of Ge on Si (100) by vacuum evaporation. Journal of Applied Physics. 1983;**54**:5466-5469

[45] Yang R et al. On the frequency dispersion of the capacitance –voltage behavior of epitaxial Ge on Si p+−n junctions. Journal of Applied Physics. 2009;**106**:074511

[46] Madelung O, Rössler U, Schulz M. Semiconductors Group IV Elements, IV-IV and III-V Compounds. Part b - Electronic, Transport, Optical and Other Properties. Springer; 2002

[47] Miller SL. Avalanche breakdown in germanium. Physics Review. 1955;**99**: 1234-1241

[48] Okuto Y, Crowell CR. Threshold energy effect on avalanche breakdown voltage in semiconductor junctions. Solid-State Electronics. 1975;**18**(2): 161-168

[49] Ando H et al. Characteristics of germanium avalanche photodiodes in the wavelength region of 1-1.6μm. IEEE Journal of Quantum Electronics. 1978; **QE-14**(11):804–809

[50] Liu J. Photonics Device. Cambridge University Press; 2005. ISBN: 9780511614255

[51] Kao K et al. Direct and indirect band-to-band Tunneling in germanium-based TFETs. IEEE Transactions on Electron Devices. 2012;**59**(2):292-301

[52] Vossen JL et al. Preparation of surfaces for high quality interface formation. Journal of Vacuum Science and Technology A. 1984;**2**(2):212-215

[53] Nayfeh A. Heteroepitaxial growth of relaxed germanium on silicon [Ph.D thesis]. Stanford University; 2006

[54] Brabant PD et al. Moisture requirements to reduce interfacial sub-oxides and lower hydrogen pre-bake temperatures for RPCVD Si epitaxy. Journal of Crystal Growth. 2013;**381**: 33-36

[55] Raynal P et al. Wet and Siconi® cleaning sequences for SiGe p-type metal oxide semiconductor channels. Microelectronics Engineering. 2018;**187–188**:84-89

[56] Haddara YM et al. Silicon-germanium: Properties, growth and applications. In: Chapter 22 of Springer

Handbook of Electronic and Photonic Materials. 04 October 2017. ISBN 978-3-319-48933-9

[57] Virgilio M, Grosso G. Type-I alignment and direct fundamental gap in SiGe based heterostructures. Journal of Physics: Condensed Matter. 2006;**18**: 1021-1031

[58] Liu J et al. Ge-on-Si optoelectronics. Thin Solid Films. 2012;**520**(8):3354-3360

[59] Teherani J et al. Extraction of large valence-band energy offsets and comparison to theoretical values for strained-Si/strained-Ge type-II heterostructures on relaxed SiGe substrates. Physical Review B. 2012;**85**: 205308

[60] Vines P et al. High performance planar germanium-on-silicon single-photon avalanche diode detectors. Nature Communications. 2019;**10**:1-8. Article number: 1086

[61] Thorburn F et al. Ge-on-Si single-photon avalanche diode detectors for short-wave infrared wavelengths. Journal of Physics: Photonics. 2022;**4**: 012001

[62] Kim G et al. Enhanced frequency response associated with negative photoconductance in an InGaAs/InAlAs avalanche photodetector. Applied Physics Letters. 2003;**83**(6):1249-1251

[63] Moise TS et al. Optically switched resonant tunneling diodes. Applied Physics Letters. 1995;**66**(9):1104-1106

[64] England P et al. Optical switching in a resonant tunneling structure. Applied Physics Letters. 1991;**58**(9):887-889

[65] Lyo I, Avouris P. Negative differential resistance on the atomic scale: Implications for atomic scale devices. Science. 1989;**245**(4924): 1369-1371

[66] Zaoui WS et al. Frequency response and bandwidth enhancement in Ge/Si avalanche photodiodes with over 840GHz gain-bandwidth-product. Optics Express. 2009;**17**(15): 12641-12649

[67] Kim G et al. NDR-effect vertical-illumination-type Ge-on-Si avalanche photodetector. Optics Letters. 2018; **43**(22):5583-5586

[68] Ning D et al. High speed waveguide-integrated Ge/Si avalanche photodetector. In: Optical Fiber Communication Conference, Paper OM3K.3. 2013

[69] Yuan Y et al. 64 Gbps PAM4 Si-Ge waveguide avalanche photodiodes with excellent temperature stability. Journal of Lightwave Technology. Light Counting Ethernet Transceivers Forecast. LightCounting. March 2020; **38**(17):4857-4866

[70] Wang G et al. A time-delay equivalent-circuit model of ultrafast p-i-n photodiodes. IEEE Transactions on Microwave Theory and Techniques. 2003;**51**(4):5583-5586

[71] Dai D et al. Equivalent circuit model of a Ge/Si avalanche photodiode. In: IEEE International Conference on Group IV Photonics. 2009

[72] Novack A et al. Germanium photodetector with 60 GHz bandwidth using inductive gain peaking. Optics Express. 2013;**21**(23):28387-28393

[73] Committee SFF. SFF-8679 specification for QSFP+ 4X. Hardware and Electrical Specification. SNIA SFF TWG technology affiliate. 2018

[74] Lu Z et al. Geiger-mode operation of Ge-on-Si avalanche photodiodes. IEEE Journal of Quantum Electronics. 2011; **47**(5):731-735

[75] Wang B et al. A compact model for Si-Ge avalanche photodiodes. In: IEEE 15th International Conference on Group IV Photonics (GFP). 2018

[76] Qi X. High frequency characterization and modeling of on-chip interconnects and RF IC wire bonds [Ph.D thesis]. Stanford University; 2001

[77] Gao J. Optoelectronic Integrated Circuit Design and Device Modeling. New York, NY, USA: Wiley; 2011

[78] Nada M et al. High-speed III-V based avalanche photodiodes for optical communications—The forefront and expanding applications. Applied Physics Letters. 2020;**116**:140502-1–140502-5

[79] Mazzini M, 'Technical feasibility of 56Gbaud PAM4 optical link budget based on experimental measurements', 400 Gb/s Ethernet Task Force, SMF Ad Hoc Conference Call. 2014

[80] Sze SM, Ng KK. Physics of Semiconductor Devices. 3rd ed 2007. ISBN: 978-0-470-06832-8

[81] IEEE/MSA 100G-ER1-40km specification

[82] Saliou F et al. Optical access network interfaces for 5G and beyond. Journal of Optical Communications and Networking. 2021;**13**(8):D32–D4

[83] Yu R, Pan D. Silicon Photonics Applications for 5G and Data Centers, Optical Fiber Communication Conference, Paper M3A.5. 2021

[84] Yu R, Chang F. Proposed 800G LR4 Baseline with PAM4 IMDD', IEEE P802.3df 200 Gb/s, 400 Gb/s, 800 Gb/s, and 1.6 Tb/s Ethernet Task Force meeting. March 2022

[85] Tian Y et al. 800Gb/s-FR4 specification and interoperability analysis. In: Optical Fiber Communications Conference and Exhibition, Paper W7F6. 2021

[86] Nagarajan R, Lyubomirsky I. Next-Gen Data Center Interconnects: The Race to 800G. COBO Webcast; 2021